LONDON MATHEMATICAL SOCIETY STUDENT TEXTS

Managing Editor: Ian J. Leary,
Mathematical Sciences, University of Southampton, UK

58 Computational algebraic geometry, HAL SCHENCK
59 Frobenius algebras and 2D topological quantum field theories, JOACHIM KOCK
60 Linear operators and linear systems, JONATHAN R. PARTINGTON
61 An introduction to noncommutative Noetherian rings (2nd Edition), K. R. GOODEARL & R. B. WARFIELD, JR
62 Topics from one-dimensional dynamics, KAREN M. BRUCKS & HENK BRUIN
63 Singular points of plane curves, C. T. C. WALL
64 A short course on Banach space theory, N. L. CAROTHERS
65 Elements of the representation theory of associative algebras I, IBRAHIM ASSEM, DANIEL SIMSON & ANDRZEJ SKOWROŃSKI
66 An introduction to sieve methods and their applications, ALINA CARMEN COJOCARU & M. RAM MURTY
67 Elliptic functions, J. V. ARMITAGE & W. F. EBERLEIN
68 Hyperbolic geometry from a local viewpoint, LINDA KEEN & NIKOLA LAKIC
69 Lectures on Kähler geometry, ANDREI MOROIANU
70 Dependence logic, JOUKU VÄÄNÄNEN
71 Elements of the representation theory of associative algebras II, DANIEL SIMSON & ANDRZEJ SKOWROŃSKI
72 Elements of the representation theory of associative algebras III, DANIEL SIMSON & ANDRZEJ SKOWROŃSKI
73 Groups, graphs and trees, JOHN MEIER
74 Representation theorems in Hardy spaces, JAVAD MASHREGHI
75 An introduction to the theory of graph spectra, DRAGOŠ CVETKOVIĆ, PETER ROWLINSON & SLOBODAN SIMIĆ
76 Number theory in the spirit of Liouville, KENNETH S. WILLIAMS
77 Lectures on profinite topics in group theory, BENJAMIN KLOPSCH, NIKOLAY NIKOLOV & CHRISTOPHER VOLL
78 Clifford algebras: An introduction, D. J. H. GARLING
79 Introduction to compact Riemann surfaces and dessins d'enfants, ERNESTO GIRONDO & GABINO GONZÁLEZ-DIEZ
80 The Riemann hypothesis for function fields, MACHIEL VAN FRANKENHUIJSEN
81 Number theory, Fourier analysis and geometric discrepancy, GIANCARLO TRAVAGLINI
82 Finite geometry and combinatorial applications, SIMEON BALL
83 The geometry of celestial mechanics, HANSJÖRG GEIGES
84 Random graphs, geometry and asymptotic structure, MICHAEL KRIVELEVICH et al
85 Fourier analysis: Part I - Theory, ADRIAN CONSTANTIN
86 Dispersive partial differential equations, M. BURAK ERDOĞAN & NIKOLAOS TZIRAKIS
87 Riemann surfaces and algebraic curves, R. CAVALIERI & E. MILES
88 Groups, languages and automata, DEREK F. HOLT, SARAH REES & CLAAS E. RÖVER
89 Analysis on Polish spaces and an introduction to optimal transportation, D. J. H. GARLING
90 The homotopy theory of $(\infty,1)$-categories, JULIA E. BERGNER
91 The block theory of finite group algebras I, M. LINCKELMANN
92 The block theory of finite group algebras II, M. LINCKELMANN
93 Semigroups of linear operators, D. APPLEBAUM
94 Introduction to approximate groups, M. C. H. TOINTON
95 Representations of finite groups of Lie type (2nd Edition), F. DIGNE & J. MICHEL
96 Tensor products of C^*-algebras and operator spaces, G. PISIER
97 Topics in cyclic theory, D. G. QUILLEN & G. BLOWER

London Mathematical Society Student Texts 98

Fast Track to Forcing

MIRNA DŽAMONJA
Associate Member
IHPST – Institute for the History and
Philosophy of Science and Technology
CNRS – Université Panthéon-Sorbonne
Paris, France

CAMBRIDGE
UNIVERSITY PRESS

CAMBRIDGE
UNIVERSITY PRESS

University Printing House, Cambridge CB2 8BS, United Kingdom

One Liberty Plaza, 20th Floor, New York, NY 10006, USA

477 Williamstown Road, Port Melbourne, VIC 3207, Australia

314–321, 3rd Floor, Plot 3, Splendor Forum, Jasola District Centre, New Delhi – 110025, India

79 Anson Road, #06–04/06, Singapore 079906

Cambridge University Press is part of the University of Cambridge.

It furthers the University's mission by disseminating knowledge in the pursuit of education, learning, and research at the highest international levels of excellence.

www.cambridge.org
Information on this title: www.cambridge.org/9781108420150
DOI: 10.1017/9781108303866

First published 2021

A catalogue record for this publication is available from the British Library.

Library of Congress Cataloging-in-Publication Data
Names: Džamonja, Mirna, 1965– author.
Title: Fast track to forcing / Mirna Džamonja.
Description: New York : Cambridge University Press, 2020. | Series: London Mathematical Society student texts ; 98 | Includes bibliographical references and index.
Identifiers: LCCN 2020018740 | ISBN 9781108420150 (hardback)
Subjects: LCSH: Forcing (Model theory) | Axiomatic set theory.
Classification: LCC QA9.7 .D93 2020 | DDC 511.3/4–dc23
LC record available at https://lccn.loc.gov/2020018740

ISBN 978-1-108-42015-0 Hardback
ISBN 978-1-108-41314-5 Paperback

Dedicated to Jean–Marc Vanden–Broeck, for his never–ending support in all my mathematical adventures.

Contents

Preface

The idea of this book grew out of the lecture notes for a first-year postgraduate module I taught for the MAGIC network of universities in Britain in 2009–2013. Part One of the book roughly corresponds to what was taught there, while the second part is new material. The MAGIC network, at the time of the writing of this book, consists of 21 UK universities and offers postgraduate courses to students enrolled for a Ph.D. in Mathematics, with the idea of providing a quick access to a subject to students who specialise in subjects away from it. The set theory module had ten one-hour lectures. The challenge, for me, was to start these students from the ground zero in set theory and present the subject up to and including the method of forcing. The module proved very popular not only with the students over the remote network but also with some unexpected listeners, such as Ms Eva Roberts, who was the excellent technology assistant responsible for the technical aspects of our video conferencing system, and who proved to be a very interested and informed member of the set theory audience! They encouraged me to make the material available to broader audience.

This book does not really aim to make a set theorist out of you, but it might happen if you read everything. If you only read the first part, it will not even make you competent enough to pass a serious first-year graduate module in set theory in a mathematical logic department. For this you will have to start by consulting one of the classic references, such as [51], [63] (this one being my personal favourite), or a more recent [48]. You will have to work through a large number of exercises and read and reproduce many proofs. There is no fast track to this. However, this book, which does not have a single exercise and skips many proofs, will help you inform yourself of this exciting area of research, unjustly considered too complex to be explained to an interested uninformed listener. If you are a mathematician, it may be that this knowledge will influence you to see some foundational aspects in your own work. If you are a budding set theorist then this book will allow you to plunge into those more serious references with confidence–they are not a particularly easy read if you have not seen any of this material before! And if you are not a mathematician at all, this book will form part of your general culture, somewhat different than the usual aspects of it - but if one can learn Marcel Proust by heart in order to cite him at parties, why not some more esoteric stuff such as foundations of mathematics!

At some point though, maybe you have read the first part of the book, cited foundations of mathematics at many parties and started being bored by the parties. You want to learn more forcing because you fell in love with it! Part Two is made for you. It will tell you much more about forcing from the time it was invented to now. It will tell you about the successes and the challenges and it will tell you about many open questions. It aims to share with you what only the cognoscenti seem to know: combinatorial set theory did not die with the invention of forcing. It was reborn. I hope that some of my colleagues in set theory will find a few interesting sentences in this part of the book, as some of it is rather new material and some not yet published.

And now for Something Completely Different

Finally, you, the reader who goes to parties and cites Proust, know that there is another reason I wrote this book. Many years ago, when I was a finishing undergraduate student in the city of Sarajevo, in what is now Bosnia and Herzegovina and then was Yugoslavia, I was supposed to produce a 4th year thesis on a topic of my choice. I chose forcing, obtained a book on forcing (in this instance, Kunen's book [63] that I recommended you above) and tried to read it on my own. It was impenetrable to me, none of my teachers knew forcing and could not help me, and I got seriously stuck on the exercises in Chapter II. Although I knew a lot of classical set theory, I did not know any logic and it was blocking me, but I did not understand that this was what was blocking me. In fact, probably the most serious problem was that I could follow the arguments line-by-line, but I did not have any intuition. At any rate, I wrote my 4th year thesis on a different topic (Category Theory) and was fortunate enough to be accepted as a graduate student at the University of Wisconsin-Madison and be supervised by Prof. Kenneth Kunen himself. His wonderful graduate course complemented his book, just as it was intended in his writing, and I finally broke through to become an insider. Since that time I have lived in New York, London and Paris and most importantly if one is a set theorist, in Jerusalem, but I have never forgotten how difficult it is to become an insider of set theory if one lives in a place where there are no others to talk to about the subject.

I like to share. I have taught logic and set theory in many places, including the unforgettable African Institute of Mathematical Sciences in Mbour, Senegal, where I had the privilege to address the students chosen from 17 different African countries, for many of whom this was the first time they had seen my subject. Seeing their smiles at the end of every lecture, knowing that we have shared a previously unknown secret, gives a unique thrill.

I wrote this book as if I were teaching to such an audience. I hope the book will help you get an intuition for set theory and I hope you will enjoy it.

I would like to gratefully acknowledge the support of the School of Mathematics at the University of East Anglia (UEA) in Norwich, UK, where I was a Professor of Mathematics at the time of writing this book, and of the scientific consulting company Logique Consult in Paris, where I am the CEO. All my thanks are due to my former Ph.D students Dr Omar Selim and Dr Francesco Parente, in chronological order, for their help with various mathematical and technical aspects of this book. Friends and students helped me with proof-reading, among them my student Dr Cristina Criste, UEA Ph.D. student Mark Kamsma and my colleagues Prof. David Buhagiar from the University of Malta and Prof. Lorenz Halbeisen from ETH in Zurich. Prof. Uri Abraham from the Ben-Gurion University of the Negev in Beer Sheva pointed out oversimplifications in a previous representation of Theorem 7.8.3. The Cambridge University Press team have been wonderful and I am really grateful to Roger Astley for his enthusiasm for the concept; Roger, Clare Dennison and Anna Scriven for various aspects of the editing process and the copy editor Jon Billam for his informed and interesting comments. Thank you all very much!

PART ONE

LET'S BE INDEPENDENT

1
Introduction

Set theory has a dual nature as a subject which is on the one hand foundational and on the other, mathematical. Its foundational background comes from the context in which after the Hilbert programme there came a search for axioms that would encapsulate the entire mathematics the way that, say, the axioms of Euclidean geometry in David Hilbert's rendering of it, encapsulate that subject. The late nineteenth century was the time when many mathematical developments had made it clear that it was no longer possible to do mathematics without being aware of the foundations behind it, as paradoxes were starting to appear. Hence the Hilbert programme specifically set the goal of finding foundations for mathematics. Set theory has soon emerged as an acceptable, if not entirely perfect answer–as we shall explain later on. The subject of set theory is probably still best known for its connections with foundations.

However, its origin and development for most of its existence have been purely mathematical. While researching properties of trigonometric series of real numbers, George Cantor in the 1870s came up with the notion of a set. This was soon followed by his discovery of the ways to produce sets out of given sets and to use sets to discuss the notion of infinity. The mathematical content of set theory is probably best summarised by stating that it is the study of infinite sets. In this book we shall discuss many aspects of infinitary combinatorics, and we shall see how these developments overlap with foundational issues, which we shall also discuss. The first part of our journey will finish in about 1963, when the independence of the Continuum Hypothesis was proved.

Life does not end in 1963 and much has happened in set theory since. The second part of the book has the ambitious goal to tell the reader something about the newer developments of the subject, sometimes quite advanced.

A final word, on history. While in the first part we might sometimes have let the reader rely on the standard references to find the exact history of the results presented, we try much harder in the second part since many of the developments in it are quite new and not in any standard books. It was important to us to be lighthearted and quick in the first part and so, much interesting history about Cantor developing set theory and obsessing with the Continuum Hypothesis, the axioms coming to life after a struggle for existence of set theory that was needed after the discovery of the Russell Paradox, Hilbert's 'Nobody will take us out of the paradise of Cantor' and many darker moments, such as Felix Hausdorff and Dénes König taking their lives rather than falling into the hands of the Nazis, is missing. The reader might get inspired to find out more about the history of set theory, it is a uniquely interesting read. An advanced reader looking for his name in the references will hopefully either be pleased or forgiving. There is too much set theory to be told in one book and we made a selection that fits our story and does not judge the rest of it.

2
Axiomatic Systems

In order to formalise mathematical thinking, Euclid's books (in the context of geometry) give the process of *axiomatisation*. This procedure first gives us a list of *objects* that are not defined but accepted as given. Then we get a list of *axioms* which are statements about these objects which are all assumed to be true. We assume also that we are endowed with the notion of *logical deduction* which we are allowed to apply to the axioms and the statements that have already been derived from the axioms using valid deduction. In this way we obtain *theorems*. The complete list of theorems that can be derived from the axioms is called a *theory*.

For example, the objects of Euclidean geometry are points, lines and planes. Notice that even though we may have a natural interpretation of these objects as the actual points, lines and planes, these objects are not defined but are taken for granted. Moreover, it is perfectly possible that there is some other interpretation of these axioms which in fact does not resemble our usual interpretation. For example, if we are only interested in the Axioms I–IV of Euclid, then an interpretation of points as points on a sphere and the lines as great circles of the sphere, gives a valid model of these axioms (this model is known as the spherical geometry). The basic objects can satisfy or not satisfy the basic relation, which is that one of *incidence* or *being on*, as in 'a point is on a line'.

The original axioms of Euclidean geometry are the Axioms I–V, where for example the fifth axiom is equivalent to stating that through every point not on a given line, there is exactly one line parallel to the given one. This particular axiom is true under the usual interpretation, but not in the spherical geometry. The axioms are used to produce theorems, such as the theorem stating that the sum of the angles of a triangle is 180 degrees (this also is not true in spherical geometry).

We shall be dealing with a specific axiomatisation of set theory, known as ZFC. These letters stand for ***Z****ermelo–****F****raenkel set theory with* **C***hoice*. This axiomatisation emerged after many years of effort (about 20) by various mathematicians, proposing different axiom schemes. It emerged as a possible answer to the Hilbert programme. Although we shall come back to this point, we do not give much historical or philosophical perspective to the development of ZFC set theory here or to set theory that existed before the development of the axioms (we have written another book on these matters, intended for a more philosophical audience, [24]). It is the best known foundation for mathematics and it is widely accepted, although far from being perfect or unique. Some of the imperfections, such as incompleteness and the independence, will be the object of study here. From now on, whenever we say 'set theory' we mean 'set theory as axiomatised by ZFC', unless we specifically say that some other axiomatisation is assumed.

In set theory, the basic objects are quite simple: there is only one kind of object, namely sets. The basic relation is also very simple, it is the membership relation $\in$. What is more complex are the axioms. There are not just finitely many axioms, but in fact infinitely many of them. In order to express them we need to have a better way of writing axioms than just quoting whole sentences, as we did in the example of Axiom V above. Axioms of set theory are presented in the context of first order logic, so let us start with an introduction to that subject.

2.1 First Order Logic

Mathematical logic is basically a mathematical language in which we can express many concepts from everyday mathematical usage. This includes, for example, the notion of a group, the notion of a field etc.

In first order logic we first fix a *language*, which consists of

- constant symbols
- relation symbols and
- function symbols.

Which language we use depends on the intended application. For example, the language of additive groups is $\mathcal{L} = \{+\}$ where $+$ is a binary function symbol. All the other symbols needed for groups (such as 0) can be defined and their properties proved from this language and the axioms. The language of graphs is $\mathcal{L} = \{R\}$ where R is a binary relation symbol.

In addition to the symbols from the language we are also allowed to use the symbols $=$, (,) and logical symbols $\wedge$, $\vee$, $\neg$, $\implies$, $\iff$ as well as the quantifiers $\forall$ and $\exists$ (in fact some of these can be derived from the others, but we shall not go into that). At our disposal we also have an infinite number of *variables* $v_0, v_1, v_2, \ldots$, or more often $x, y, z, \ldots$ From these symbols we are allowed to make *terms*:

- each variable is a term, each constant is a term;
- if f is an n-ary function symbol and $\tau_1, \ldots, \tau_n$ are terms, then so is

$$f(\tau_1, \ldots, \tau_n);$$

- each term is obtained by a finite number of applications of these rules.

From terms we can make *formulas* as follows:

- for any n-ary relation symbol R and terms $\tau_1, \ldots, \tau_n$, the expression

$$R(\tau_1, \ldots, \tau_n)$$

  is a formula (including the terms themselves, which may be considered as $\tau = R(\tau)$ for the trivial unary relational symbol R whose interpretation is $R(x)$ iff $x = x$);
- if φ, φ' are formulas so are $\neg\varphi$, $(\varphi \wedge \varphi')$, $(\varphi \vee \varphi')$, $(\varphi \implies \varphi')$, $(\varphi \iff \varphi')$;
- if φ is a formula and x a variable then $(\exists x)\,\varphi(x)$ and $(\forall x)\,\varphi(x)$ are formulas;
- each formula is obtained by a finite number of applications of these rules.

These rules imply that every formula of first order logic has a finite number of negations, disjunctions, conjunctions, implications and equivalences, and a finite number of quantifiers. In fact, some of these operations can be defined from the others, for example $\varphi \implies \psi$ is equivalent to $\psi \vee \neg\varphi$. In a minimalist definition, it suffices to have the operations $\neg$ and $\wedge$, and the quantifier $\exists$.

Note that there are infinitely many formulas in any non-empty language, because there are infinitely many variables allowed.

2.2 Language of Set Theory

The language of set theory is $\mathcal{L} = \{\in\}$ where $\in$ is a binary relation symbol interpreted as the 'membership relation'. As we shall see in the sequel, all set-theoretic symbols we know, such as $\subseteq$, $\cap$ or $\cup$, can be derived from this language and the axioms of set theory.

3
Zermelo–Fraenkel Axioms and the Axiom of Choice

Every axiomatic system has its *basic objects and relations* and the *axioms* that describe these objects. Let us go back to the example of Euclidean geometry and develop further what we said about it in the previous chapter: the basic objects of Euclidean geometry are points, lines and planes. They are not defined. The basic relationship between these objects is the relation of incidence or *membership*. There are five axioms, for example:

$$\text{Through every two distinct points there is a line} \tag{I}$$

In the context of first order logic, we could write the axioms using the language $\mathcal{L} = \{p, l, P, \in\}$ where p, l, P are unary relations symbols (whose intended meaning is 'is a point', 'is a line', 'is a plane') and $\in$ is a binary symbol (whose intended meaning is 'belongs to'). Then we have

$$\forall x_1, x_2 (p(x_1) \wedge p(x_2) \wedge \neg(x_1 = x_2) \implies (\exists x_3)(l(x_3) \wedge x_1 \in x_3 \wedge x_2 \in x_3)) \tag{IA}$$

In our mathematical discourse we most often do not write things in such gory detail, but the point of foundations is to know for sure that we could have written everything we do on this level of detail if we had wanted to.

3.1 Axioms of Set Theory

The *basic objects* of set theory are sets and they are not defined. All that we are allowed to assume about this undefined notion of a set is contained in the following list of axioms, which are collectively known as the axioms of ZFC (Zermelo–Fraenkel with Choice). Most of the axioms are natural and when interpreted correctly, will certainly be familiar from everyday mathematical discourse. Nevertheless, let us point out that there are other axiomatic systems for set theory, sometimes studied in logic (although rather rarely in mathematics). We shall not discuss any of them here, but a discussion about this can be found in [63].

We first claim that there are objects to be considered.

Axiom 0 There is a set $(\exists x)\,(x = x)$.

This is reasonable, but in fact this axiom follows from the basic logic premise that the domain of discourse is not empty. We only put it here to start our discussion.

Next we must consider what it means for two sets to be equal. Recall that in our everyday work when we wish to show that the sets X and Y are equal we usually proceed as follows:

Let $z \in X$ so that... and so $z \in Y$. Now let $z \in Y$ so that... and so $z \in X$. Therefore $X = Y$.

The 'Therefore' in the above argument needs to be asserted.

Axiom of Extensionality Two sets have exactly the same elements if and only if they are the same set or

$$\forall X \forall Y\,(\forall z(z \in X \iff z \in Y) \implies X = Y).$$

The Axiom of Extensionality tells us that there is nothing to a set but its elements: a set is determined by its elements. So if we intuitively consider a set as a box of objects then it should not matter to us what the colour or shape of this box is; what matters is what's inside the box, and nothing else.[1]

In the above axiom we used the notation familiar from the usual mathematical discourse, when we denote sets by capital letters and the elements of these sets by lower case letters. In fact, this distinction does not make much sense in set theory, because we have already declared that all our objects are sets, so the elements of sets are sets themselves. For this reason we shall not often make the lower case-upper case distinction, and we start applying that rule straight away.

Axiom of Pairing For every two sets x, y there is a set whose elements are exactly x, y or

$$\forall x \forall y \exists z (x \in z \land y \in z \land \forall w (w \in z \implies w = x \lor w = y)).$$

[1] 'It's what's inside that counts!'

Given sets x and y this axiom states the existence of an unordered pair $z = \{x, y\}$. The Axiom of Pairing looks too innocent to be worth our while, but with it we can immediately give a formal definition of a notion that we have seen many times before, that is, the notion of an ordered pair (x, y). For any sets x, y we let $(x, y) \stackrel{\text{def}}{=} \{\{x\}, \{x, y\}\}$, that is we *declare by definition* that this is what is meant by (x, y). We are justified in doing so, because the Axiom of Pairing has given us the right to form the set on the right hand side, and we simply decide to call it (x, y). It might seem that this definition of an ordered pair is somewhat arbitrary or ad hoc. However, what is important is that this definition most definitely encapsulates the following idea: for all sets a, b, c and d, we have

$$(a, b) \neq (c, d) \iff a \neq c \vee b \neq d. \tag{3.1}$$

Any other definition of an ordered pair that satisfies (3.1) would also do, for example we could have set $(x, y) \stackrel{\text{def}}{=} \{\{y\}, \{x, y\}\}$! But we did not and the reader is asked to use the definition of an ordered pair given above. As we shall see below, the notion of an ordered pair will allow us to define another very familiar object: a function.

Next consider a set, say A, of sets. It is often the case that we wish to take the union of all its objects, that is the set whose elements are exactly those objects that are an element of one of the elements of A. Luckily, we shall now introduce a notation to not have to write such a description every time! We might write something like (note the unary $\bigcup$ symbol for the union of a large family of sets):

$$\text{Let } Z = \textstyle\bigcup_{a \in A} a.$$

The assertion that we can always do this and still consider Z a set is our next axiom. (We have gone to the capitalisation notation again to improve the readability.)

Union Set Axiom $(\forall A)(\exists Z)(\forall b)(b \in Z \iff (\exists a \in A)(b \in a)).$

The following axiom is not a single axiom, but a scheme involving infinitely many axioms: one for every formula of the language of set theory. It is in this axiom (and in the Comprehension Scheme below) that we take the benefit of having used a logical formalisation: it is because of this formalisation that we are able to express infinitely many axioms in one statement, as we shall now see.

To formulate the next axiom we introduce a shorthand: let $\varphi(x, y, x_1, \ldots, x_n)$ be a formula. Then for any choice of $y, x_1, \ldots, x_n$ the notation

$$(\exists! x)\ \varphi(x, y, x_1, \ldots, x_n)$$

is a replacement for

$$(\exists x)(\varphi(x, y, x_1, \ldots, x_n) \wedge (\forall z)(\varphi(z, y, x_1, \ldots, x_n) \implies z = x)).$$

We also use the convention that when writing $\varphi(x, y, x_1, \ldots, x_n)$, we mean that the free variables of φ are among $\{x, y, x_1, \ldots, x_n\}$, but do not necessarily include them all.

Replacement Axiom Scheme For each formula $\varphi(x, y, x_1, \ldots, x_n)$ we have

$$(\forall x_1, \ldots, x_n, I)((\forall x \in I)(\exists! y\, \varphi(x, y, x_1, \ldots, x_n) \implies$$

$$(\exists A)(\forall z)(z \in A \iff (\exists x \in I)\varphi(x, z, x_1, \ldots, x_n))).$$

Let us pause for a moment to consider this hmm, monster of a statement. Loosely, this axiom is saying that if φ is an 'assignment' whose 'domain' is a *set* I then the range of φ, A, is also a set. Here is an argument where this axiom is used and one which the reader has probably seen somewhere before:

Let I be a set and for each $i \in I$ let A_i be the set... Now let $Z = \bigcup_{i \in I} A_i$.

To justify the existence of the set Z we first use the axiom of replacement to assert that the set $A = \{A_i : i \in I\}$ exists. Then by the axiom of union we have $Z = \bigcup A$.

We now come to another axiom scheme, whose formulation took a long while and corresponded to the resolution of a huge crisis in set theory. The crisis was the discovery of a paradox by Bertrand Russell in [95], in fact a contradiction that Russell found in Gottlob Frege's theory of predicates but which applies to Naive (this means non-axiomatised) Set Theory as well. The Comprehension Axiom Scheme was proposed by Ernst Zermelo in [127], and it resolved the paradox. *The Russell Paradox* is the following: assume that given any formula φ, we can form the set of all sets x such that $\varphi(x)$ holds. Then the following is a set: $\{x : x \notin x\}$. Let us call that set B. Now ask if $B \in B$. If the answer to this question is either yes or no, we get a contradiction. The paradox shows that we cannot have an unrestricted freedom on how we form sets using a given formula, as had been done by early set theorists. We can, however, given a set y and a formula φ, form the set of all elements of y which satisfy the formula φ:

Comprehension Axiom Scheme For each formula $\varphi(x, y, x_1, \ldots, x_n)$ we have

$$(\forall x_1, \ldots, x_n, y)(\exists z)(\forall x)(x \in z \iff x \in y \wedge \varphi(x, y, x_1, \ldots, x_n)).$$

It is a good idea for the reader to check at this point that the Comprehension Axiom Scheme does not lead to the contradiction exemplified by the Russell paradox. A version of Russell's paradox can also be used to see that the family of all sets does not form a set.

Comprehension is used very often in mathematics. For example if we assume for the moment that the set of positive integers numbers $\mathbb{N}^+ = \{1, 2, 3, \ldots\}$ exists (we cannot prove its existence only on the basis of the axioms we have set so far), then defining the set of even positive numbers is an instance of Comprehension. Indeed we let $\varphi(n, \mathbb{N}^+)$ say that 'n is an even positive number'. Then Comprehension asserts that the collection $\{n \in \mathbb{N}^+ : n \text{ is even}\}$ exists.

As we mentioned before, the last two axioms are not really *two* axioms. Actually each of them is an infinite collection of axioms (axiom scheme), each of which asserts something different depending on the choice of φ. We shall not develop this point further, but it is worthwhile stating that in fact one can prove that no finite list of axioms for *ZFC* has the same power as the full list of axioms. That is to say, *ZFC* is not *finitely axiomatisable*, see the work of Richard Montague [81] and Corollary 7.4.4.

Before we move on to the remaining axioms we can use the ones we have given so far to set up some familiar notation. We can define a relation $\subseteq$ between sets:

$$x \subseteq y \stackrel{\text{def}}{=} (\forall z)(z \in x \implies z \in y).$$

Now let x be a set – at least one exists by Axiom 0 - and by Comprehension let $\emptyset$ be the unique (by Extensionality) set such that

$$\emptyset = \{y \in x : \neg(y = y)\}.$$

We call this set (of course) the *empty set*. Given any set define the *successor* of x, $S(x)$ to be the unique set $x \cup \{x\}$. As an exercise one can verify that the above axioms justify such a definition. We can now define the *natural numbers* (although we cannot yet define the set of natural numbers). Namely, we identify 0 with $\emptyset$, 1 with the set $\{\emptyset\} = \{0\}$ (hence $1 = S(0)$), 2 with the set $\{0, 1\}$ (hence $2 = S(1)$), etc. In general, n is a natural number if it is obtained by a successive application of the S operation on the set $\emptyset$ finitely many times.

Given sets x and y define the set

$$x \cap y = \{z \in x : z \in y\}.$$

This makes sense by Comprehension and Extensionality. In fact, we can do better than this: given any $F \neq \emptyset$, we can define $\bigcap F = \{z : (\exists x \in F)(z \in x)\}$. Notice that this makes sense by Comprehension, but it would not have made sense had we not assumed that $F \neq \emptyset$.

Some other notions we can define on the basis of the axioms so far are e.g. the *Cartesian product* of two sets X and Y and the notion of a relation and a function. As we would expect we let

$$X \times Y = \{(x, y) : x \in X \wedge y \in Y\}.$$

This definition however requires a bit more thought than that needed for the definitions of ordered pair, successor and intersection. In order to illustrate that already at this level things can get a bit complicated, let us try to convince the reader that our definition of the Cartesian product makes sense. We have already seen that for any two sets x and y the set (x, y) exists and is unique, so in particular we have that for any set y

$$(\forall x \in X)(\exists! z)\, z = (x, y).$$

Thus for each set y, and so in particular for each set $y \in Y$, we have the assignment $X \ni x \mapsto (x, y)$. By Replacement we may form the set

$$X \times \{y\} = \{(x, y) : x \in X\}.$$

This set is again unique so applying Replacement again[2] we may form the set

$$\{X \times \{y\} : y \in Y\}.$$

Now by Union

$$X \times Y = \bigcup \{X \times \{y\} : y \in Y\}.$$

Let us carry on with the definitions. A *relation* between the elements of a set X and a set Y is any subset of $X \times Y$. A *function* from X to Y is a relation f such that for any $x \in X$ there is exactly one $y \in Y$ such that $(x, y) \in f$. Similarly, we can define an *injection*, *surjection* and *bijection* (the reader may like to do this as an exercise). We can also define when a set is *finite*: it is the case when there is a bijection between that set and a natural number.

[2] This time the we consider the assignment $y \mapsto X \times \{y\}$

We now continue with the axioms. We mentioned in the introduction that one of the main goals of set theory was to understand infinity, but based on the axioms presented so far, an infinite set need not even exist. So we assert the following.

Axiom of Infinity There exists an infinite set, which we formalise as

$$(\exists x)(\emptyset \in x \wedge (\forall y \in x)\, S(y) \in x).$$

On the basis of our definition of natural numbers, this says that there is a set that contains all natural numbers (it does not exactly say that there is a set which consists exactly of the natural numbers; as an exercise the reader may want to justify the existence of the set of natural numbers on the basis of the axioms given so far). It is a good exercise to show that such a set cannot be finite, using the above definition of finite. Hence this axiom gives us indeed the existence of an infinite set.

We use ω to denote the set of natural numbers.

Of course, given a set we should always be able to consider its power set, and it is not something that we can do on the basis of the axioms given so far. We need an extra axiom:

Power Set Axiom For every x there is y whose elements are exactly those z which are a subset of x or

$$(\forall x)(\exists y)(\forall z)\ (z \subseteq x \iff z \in y).$$

On the face of it, the Power Set Axiom is just another axiom that gives us a way of creating a new set from an old one: given x we create $y = \{z : z \subseteq x\}$. It is certainly useful in everyday mathematics. However, the Power Set Axiom, like the previous axiom, is also closely related to our pursuit of infinity. The Axiom of Infinity explicitly tells us that infinite sets exist while, by a result of Cantor in Theorem 5.0.4, the Power Set Axiom tells us that among the infinite sets – which we are now allowed to consider – there can be many *different* infinities.

We have seen that there are sets whose elements are also sets, in fact all the concrete sets that we have discussed had the property that their elements were built from the empty set by various operations. The following axiom tells us, somehow surprisingly, that we may as well assume that every possible set is built from the empty set (this implication of the axiom may not be obvious to the reader at this point, but it will be dealt with in detail when we start discussing the universe of all sets in §7.1). It is not an axiom that one would question very much or even come across when working in mathematics outside of the subject of foundations, but it is a very useful axiom for foundations. It is really due to this axiom that we are able to put some structure to the vast universe of all sets (which according to the Russell paradox, does not itself form a set).

Axiom of Foundation Any non-empty set is disjoint from at least one of its members or

$$\forall x(\neg(x = \emptyset) \implies \exists y(y \in x \wedge y \cap x = \emptyset)).$$

As a consequence of the Axiom of Foundation no set can be a member of itself. Indeed assuming foundation, assume that there is a set x such that $x \in x$. Then the set $\{x\}$, which exists by Pairing, contradicts Foundation.

The Axiom of Foundation, in fact, tells us that the membership relation orders the universe of all sets in a rather nice way. To formulate this we need the following definition.

Definition 3.1.1 We call the ordered pair (X, R), where R is a binary relation, a *total, or a linear, ordering* if and only if for every $x, y, z \in X$ the following hold (writing xRy for $(x, y) \in R$):

- (Transitivity) If xRy and yRz then xRz;
- (Irreflexivity) $(x, x) \notin R$;
- (Trichotomy) $x = y$ or xRy or yRx.

If in addition we have:

- For every non-empty subset Y of X, there exists $y \in Y$ such that for each $x \in Y$ it is not the case that xRy,[3]

then we call (X, R) a *well ordering*.

Also useful is a *partial ordering*, which satisfies all requirements of a total ordering, with the exception of the requirement of Trichotomy.

We can check that the usual ordering of the set of the natural numbers is a well ordering, while the usual ordering of the set of the real numbers is not. Foundation says that $\in$ well orders the universe of all transitive sets (namely those on which the membership relation is transitive). Our final axiom states that *any* set can be well ordered.

[3] Or $\forall Y \subseteq X(Y \neq \emptyset \implies \exists y \in Y \forall x \in Y(\neg((x, y) \in R)))$ or every non-empty subset of X has an R-minimal element.

Axiom of Choice For every set X there exists a set R such that (X, R) is a well ordering.

The above formulation is the most useful one for set theory, but it does not make it clear what the word 'choice' refers to. In fact there is an equivalent statement of the Axiom of Choice, which says that for any set $\mathcal{F}$ of non-empty sets, there is a function f with domain $\mathcal{F}$ such that for any element A of $\mathcal{F}$, the value $f(A)$ is some element of A. For finite sets $\mathcal{F}$ this axiom follows from the rest of the axioms, but it is not the case for the infinite $\mathcal{F}$. As self-evident as this axiom seems to be at first, it actually is not. Axiom of Choice has many equivalents and consequences in mathematics, some of which are somewhat controversial. The consequences include the existence of a non-Lebesgue measurable subset of the reals ([124]) and a partition of the unit sphere in $\mathbb{R}^3$ into a finite number of pieces which can be reassembled into two copies of the unit sphere (this is the so called *Banach–Tarski paradox*, see [125]). For this and other reasons the Axiom of Choice caused much controversy amongst mathematicians, perhaps because of its highly non-constructive nature: the Axiom of Choice simply asserts that a well ordering always exists but gives absolutely no hint as to how these well orderings can be constructed.[4] Even the Axiom of Infinity gives an explicit description of the set it claims to exist and similarly with the Power Set Axiom. Controversial or not, the Axiom of Choice, which has many guises, turns up everywhere in mathematics and therefore we accept it as part of foundations. For example, the Axiom of Choice is used to show that every vector space has a basis.

The following equivalent of the Axiom of Choice is very useful in set theory:

Theorem 3.1.2 (Kuratowski–Zorn Lemma) *Every non-empty partially ordered set in which every chain (i.e. a linearly ordered subset) has an upper bound, contains at least one maximal element.*

We could at this point go on and show that these axioms allow us to define the familiar objects in mathematics such as $\mathbb{Q}$ and $\mathbb{R}$ and justify the proof methods such as induction. We shall not develop these points now, although we may return to some of them as we need them.

[4] We invite the reader to try to well order the set of real numbers $\mathbb{R}$ without using the Axiom of Choice!

4
Well Orderings and Ordinals

We start with a lemma that gives an equivalent definition of a well ordering, and which is often used in practice when we check whether a set is well ordered or not. We use the notation $(L, <)$ of the strict order and $(L, \leq)$ of the order which is not necessarily strict, interchangeably, since each of these objects is induced by the other. All theorems in the theory of order translate between these versions in a straightforward manner.

Lemma 4.0.1 *Suppose that $(L, <)$ is a linearly ordered set. Then $<$ is a well order on L iff there is no infinite $<$-decreasing sequence $a_0 > a_1 > a_2 > \cdots > a_n > \ldots$.*

Proof Suppose that $<$ is a well order but that $a_0 > a_1 > a_2 > \cdots > a_n > \ldots$ is an infinite decreasing sequence. Then the subset $\{a_n : n < \omega\}$ of L does not have a $<$-least element, a contradiction.

In the other direction, suppose that the order is linear but not a well order. Then there is a non-empty subset A of L which has no $<$-least element. By induction on n we choose $a_n \in A$. Namely, let a_0 be any element of A. Since a_0 is not the least element of A, we can choose a_1 which is smaller than a_0. But, this is still not the least element, so we can choose a_2 which is still smaller etc. Given $a_n \in A$ we choose $a_{n+1} < a_n$. At the end we have obtained an infinite decreasing sequence. $\bigstar_{4.0.1}$

The reader may stop to think why we are allowed to use the method of mathematical induction in the above argument. In fact the reason is exactly because the set of natural numbers is well ordered. The axioms of Peano Arithmetic (PA) follow from the ZFC Axioms, see Chapter 6.

Definition 4.0.2 1. A set x is *transitive* iff the relation $\in$ is transitive on x, so for every $y \in x$ and $z \in y$ we have $z \in x$. In other words, every element of x is a subset of x.

2. An *ordinal* is a transitive set which is well ordered by $\in$.

Lemma 4.0.3 *If x is an ordinal, then so is $S(x)$.*

Proof If $y \in S(x)$ then either $y \in x$ so $y \subseteq x \subseteq S(x)$, or $y = x \subseteq S(x)$. Hence $S(x)$ is transitive. This easily implies that the relation $\in$ is a linear order on $S(x)$. To verify the remaining property being well ordered by $\in$, we shall give two proofs:

Proof 1: Let us use directly the definition of being a well order. So suppose that $A \neq \emptyset$ is a subset of $S(x)$. If $A \cap x \neq \emptyset$ then, since x is well ordered by $\in$ there is the $\in$-least element y of $A \cap x$. In particular $y \in x$, so y is smaller than x, and hence y is the least element of A. If $A \cap x = \emptyset$ then A must be equal to $\{x\}$, and then it clearly has the least element, namely x itself.

Proof 2: Let us use the equivalent definition of a well order coming from Lemma 4.0.1. Suppose that $(a_n)_n$ is an infinite $\in$-decreasing sequence in $S(x)$. Since there can be at most one n such that $a_n = x$, we have that $\langle a_n : a_n \neq x\rangle$ is an infinite $\in$-decreasing sequence in x, contradicting the fact that x is an ordinal. $\bigstar_{4.0.3}$

Example 4.0.4 1. $\emptyset$ is an ordinal, every $n \in \omega$ is an ordinal, ω is an ordinal.

2. $\{1, 2\}$ is not a transitive set as $0 \in 1$ but $0 \notin \{1, 2\}$.
3. The following is a well order on the set ω: define $x <^* y$ iff either x, y are both even and $x < y$, or x, y are both odd and $x < y$, or x is even and y is odd.

One of the most important properties of ordinals is that they are comparable to each other in a rather direct manner. To say this correctly, let us introduce the following general notation:

Notation 4.0.5 1. Suppose that $(L, \leq)$ is a linear order and $x \in L$. Then $\mathrm{pred}(L, x)_< = \{y \in L : y < x\}$. If $<$ is clear from the context then we omit it from the notation.

2. If $(L_1, \leq_1)$, $(L_2, \leq_2)$ are two linear orders then an *order isomorphism* between L_1 and L_2 is a bijection $f\colon L_1 \to L_2$ which preserves the order i.e. $x \leq_1 y \iff f(x) \leq_2 f(y)$. If there is such an isomorphism we write $L_1 \simeq L_2$.
3. A subset A of a linear order L is an *initial segment* if

$$(\forall b \in A)(\forall a \leq b)\, a \in A.$$

Lemma 4.0.6 *Suppose that $(A, <)$ is a well ordered set and B is a subset of A, then the restriction $(B, < \restriction B)$ is well ordered.*

Proof It is known and easy to check that $<$ is still a linear order on B. If C is a non-empty subset of B, it is in particular a non-empty subset of A, so it has the $<$-least element in A, and consequently in B. $\bigstar_{4.0.6}$

The following is a celebrated theorem of Cantor.

Lemma 4.0.7 (Cantor's Trichotomy) *Let $(L_l, \leq_l)$ for $l = 1, 2$ be two well orders. Then exactly one of the following holds:*

- *L_1 is order isomorphic to $\text{pred}(L_2, x)$ for some $x \in L_2$,*
- *L_2 is order isomorphic to $\text{pred}(L_1, x)$ for some $x \in L_1$, or*
- *L_1 is order isomorphic to L_2.*

Proof Let

$$f = \{(v, w) \in L_1 \times L_2 : \text{pred}(L_1, v) \simeq \text{pred}(L_2, w)\}.$$

Then f is a relation between the elements of L_1 and those of L_2. We claim it is a function, since if $\text{pred}(L_1, v) \simeq \text{pred}(L_2, w_1)$ and $\text{pred}(L_1, v) \simeq \text{pred}(L_2, w_2)$ then by composition $\text{pred}(L_2, w_1) \simeq \text{pred}(L_2, w_2)$. Let this be exemplified by an order isomorphism g. Suppose that $w_1 \neq w_2$, so either $w_1 <_2 w_2$ or vice versa. Say, $w_1 <_2 w_2$. Hence for some $z <_2 w_1$ we have $g(z) = w_1$ and therefore $A = \{z <_2 w_1 : z <_2 g(z)\}$ is non-empty. It has a $<_2$-minimal element, say z^*. Since $z^* <_2 w_1 <_2 w_2$, we have that for some $u <_2 w_1$, $g(u) = z^*$. If $u <_2 z^*$ then $u \in A$, a contradiction with the choice of z^*. If $u = z^*$ we get $g(z^*) = z^*$, again a contradiction. Hence $u >_2 z^*$, so $z^* = g(u) >_2 g(z^*)$, a contradiction. Hence $w_1 = w_2$.

In a similar way we can show that f is 1–1. Let $A = \{v : (\exists w)\,(v, w) \in f\}$, so A is the domain of f, and let B be the similarly defined range of f. We claim that A is an initial segment of L_1: if $u <_1 v$ and v is in A, this means that there is an order isomorphism g from $\text{pred}(L_1, v)$ onto some $\text{pred}(L_2, w)$. In particular, $g(u) = z$ is well defined. It is easy to check that $g \restriction \text{pred}(L_1, u)$ is an order isomorphism and that $(u, z) \in f$. Hence $u \in A$. In a similar way we can show that B is an initial segment.

If at least one among $A = L_1$ or $B = L_2$ is true, we are done. Suppose for contradiction that this is not the case, and let l_i for $i = 1, 2$ be the $<_{L_l}$-least element of L_l which is not in A (or in B). Then f shows that (l_1, l_2) should be in f, a contradiction.

We can similarly check that no two of the above possibilities can hold simultaneously. $\bigstar_{4.0.7}$

Theorem 4.0.8 *These are the basic properties of ordinals:*

1. *If x is an ordinal and $y \in x$ then y is an ordinal and $y = \text{pred}(x, y)$.*

2. *If x, y are ordinals and $x \simeq y$, then $x = y$.*
3. *(Trichotomy) If x, y are ordinals then exactly one of the following is true: $x \in y$, $y \in x$, or $x = y$.*
4. *For no ordinal x can we have that $x \in x$.*
5. *If x, y, z are ordinals, $x \in y$ and $y \in z$, then $x \in z$.*
6. *Any non-empty set of ordinals has the $\in$-least element.*
7. *If A is a transitive set of ordinals, then A is an ordinal.*

Proof (1) Since $y \in x$ we have $y \subseteq x$. By Lemma 4.0.6 we have that $\in$ is a well order on y, so we need to check that y is transitive. If $z \in y$, then $z \in x$ since $y \subseteq x$, and hence $z \subseteq x$. So if $w \in z$ then $w, y \in x$ so w and y must be comparable by $\in$ as $\in$ is a linear order on x. If $y \in w$ then $\{z, y, w\}$ is a non-empty subset of x with no $\in$-minimal element, a contradiction. Similarly, if $y = w$, then $\{z, y\}$ is such a subset, again a contradiction. Hence $w \in y$. This shows that y is an ordinal. Finally, by transitivity

$$\text{pred}(y, x) = \{z \in x : z \in y\} = \{z : z \in y\} = y.$$

(2) Let $f\colon x \to y$ be an order isomorphism. We claim that f is the identity, hence $x = y$. Suppose for contradiction that this is not the case and consider $\{z \in x : f(z) \neq z\}$, which is now assumed to be non-empty. Let z be its $\in$-minimal element and let $w = f(z)$. Then $f \restriction z\colon z \to w$ is an order isomorphism (as can be checked by using (1)), so for every u we have $u \in z \iff f(u) = u \in w$. By Extensionality, $z = w$, a contradiction. Hence f is the identity and using Extensionality again we obtain that $x = y$.

(3) This follows from (2) and Lemma 4.0.7.

(4) Suppose that $x \in x$ for some ordinal x. Since Lemma 4.0.3 states that $S(x)$ is an ordinal, we can obtain a contradiction by considering the $\in$-least element of the non-empty subset $\{x\}$ of $S(x)$.

(5) Notice that $x, y, z \in S(z)$ so the conclusion follows from the linearity of $\in$ in $S(z)$.

(6) Since $C \neq \emptyset$, let x be any element of C. If x is the $\in$-minimal element of C then we are done. Otherwise, let $y \in x$ be the $\in$-minimal element of $x \cap C$. Let $z \in C$. If $z \in y$ then $z \in x$, leading to a contradiction. Thus $y \in z$, or $y = z$ showing that y is the $\in$-minimal element of C.

(7) This follows from (5), (6). $\bigstar_{4.0.8}$

By the trichotomy of $\in$ from Theorem 4.0.8 we are justified in the following notation.

Definition 4.0.9 For ordinals x and y we say $x < y$ if and only if $x \in y$.

From Theorem 4.0.8 it follows that there is no set of all ordinals, by a paradox similar to Russell's paradox. Namely, if O were to be the set of all ordinals, it would follow from 4.0.8 that O itself is an ordinal, and then we would obtain a contradiction because $O \in O$. This is sometimes referred to as *the Burali-Forti paradox*, discovered by Cesare Burali-Forti in [10].[1]

Given any formula $\varphi(x)$ we may form the *class* $\{x : \varphi(x)\}$, this is the collection of all sets x such that $\varphi(x)$ holds. But as we have just seen not every class can be a set. A class which is not a set is called a *proper class*.

Another important property of ordinals is that they code all well orders:

Theorem 4.0.10 *Any well ordering (W, R) is order isomorphic to a unique ordinal.*

Proof Uniqueness follows from Theorem 4.0.8 (2). For the existence we consider the set

$$B = \{b \in W : \mathrm{pred}(W, b) \simeq \beta \text{ for some ordinal } \beta\}.$$

For each $b \in B$ let $f(b)$ be the unique ordinal such that $\mathrm{pred}(W, b) \simeq f(b)$ – such an ordinal exists by definition of B and it is unique by Theorem 4.0.8 (2). Let

$$C = \{f(b) : b \in B\};$$

this set exists by replacement.

Claim 4.0.11 C is an ordinal.

Proof of Claim 4.0.11 Since C is a set of ordinals, by Theorem 4.0.8 (7), it is enough to show that C is transitive with respect to $\in$. To this end let $\beta \in C$. Then $\beta = f(b)$ for some $b \in B$ and there exists an order isomorphism $g\colon \mathrm{pred}(W, b) \simeq \beta$. Now suppose that $\gamma \in \beta$, we want to show that $\gamma \in C$. Since g is onto and $\gamma \in \beta$, there exists $a \in \mathrm{pred}(W, b)$ such that $g(a) = \gamma$. Since g is an order isomorphism $g \restriction \mathrm{pred}(W, a)\colon \mathrm{pred}(W, a) \to \gamma$ is also an order isomorphism and so by definition $\gamma \in C$. $\bigstar_{4.0.11}$

We leave it as an exercise to show that the function

$$f\colon B \to C : b \mapsto f(b)$$

is an order isomorphism. Thus the theorem will be proved if we can show that $B = W$. Assume for a contradiction that $B \neq W$ so that $W \setminus B \neq \emptyset$. Let $b = \min(W \setminus B)$. By minimality of b if cRb then $c \in B$ and so $\mathrm{pred}(W, c) \simeq f(c)$. Thus

[1] In fact, Burali-Forti did not exactly discover this paradox: it was only some years later that several logicians, including Russell, realised that Burali-Forti had unveiled a paradox.

$\text{pred}(W, b) \simeq \{f(c) : cRb\}$ (exercise) and $\{f(c) : cRb\}$ is an ordinal (exercise) so that by definition $b \in B$, which is the desired contradiction. ★4.0.10

With this last theorem in mind, the following definition makes sense.

Definition 4.0.12 If (W, R) is a well ordering then we let $\text{otp}(W, R)$, the *order type* of (W, R), be the unique ordinal α such that $(\alpha, \in) \simeq (W, R)$.

4.1 Arithmetic Properties of Ordinals

Recall the successor operator $S(x) = x \cup \{x\}$ for any set x. We consider this operation for ordinals in light of Definition 4.0.9.

Definition 4.1.1 For any ordinal α the *ordinal successor* of α is $S(\alpha) = \alpha \cup \{\alpha\}$, which we also denote by $\alpha + 1$.

Note that given any ordinal α, there is no ordinal β such that $\alpha < \beta < S(\alpha)$. Indeed, assume for a contradiction that we could find ordinals α and β such that $\alpha < \beta < S(\alpha)$. Then, since $\beta \in S(\alpha) = \alpha \cup \{\alpha\}$, either $\beta \in \alpha$ or $\beta = \alpha$. But, since $\alpha \in \beta$, this contradicts Theorem 4.0.8 (3). Note also that by the successor operation there can be no largest ordinal: if α were the largest ordinal then the set $S(\alpha)$ would contain all the ordinals. By Comprehension we could then form the set of all ordinals which is impossible (recall the Burali-Forti paradox).

The notation '+1' is chosen to complement the notation used for the natural numbers, indeed we have for $n \geq 1$:

$$n = \{0, 1, \ldots, n-1\}, \text{ and}$$
$$n + 1 = \{0, 1, \ldots, n\} = \{0, 1, \ldots, n-1\} \cup \{n\} = n \cup \{n\} = S(n).$$

Clearly $0 \neq S(\alpha)$ for any ordinal α and also $\omega \neq S(\alpha)$ for any α since if $\omega = S(\alpha)$ for some ordinal α, we must have $\alpha \in \omega$. But since if we add one to a natural number we still get a natural number we have $\alpha < \alpha + 1 < \omega$ which as we have just shown above is impossible. These two examples suggest the following definition.

Definition 4.1.2 An ordinal α is a *successor ordinal* if and only if $\alpha = \beta + 1$ for some ordinal β. Otherwise α is called a *limit ordinal.*

Note that 0 is a limit ordinal, by definition, but the word 'limit' is really meant to be indicative of how we construct non-zero limit ordinals. For example we could (loosely) write

$$\omega = \lim_{n \to \infty} (n + 1).$$

But set theory allows us to keep going!

$$0,\ 1,\ 2, \ldots,\ \omega,\ \omega+1,\ \omega+2,\ \omega+3, \ldots,\ \omega+\omega, \ldots$$

where we write $\omega + 3$ to mean $((\omega + 1) + 1) + 1$ and so on. $\omega + \omega$ is the limit of all $\omega + 1$, $\omega + 2$, etc. We shall give some other definitions of $\omega + \omega$ below.

Observe that an ordinal α is a limit ordinal if and only if it does not have a maximal element (exercise).

Let us show how the same set can be isomorphic to different ordinals. Consider the set $\mathbb{N} = \{0, 1, 2, \ldots\}$.

- Consider the identity map $I\colon \mathbb{N} \to \omega$ defined by $I(n) = n$ and define the order $<_1 \subseteq \mathbb{N} \times \mathbb{N}$ by $n <_1 m$ if and only if $n \in m$. Then $(\mathbb{N}, <_1) \simeq (\omega, \in)$.
- Consider the map $f\colon \mathbb{N} \to \omega + 1$ defined by:

$$f(n) = \begin{cases} \omega & \text{if } n = 0 \\ n - 1 & \text{if } n \geq 1. \end{cases}$$

  Now define the relation $<_2 \subseteq \mathbb{N} \times \mathbb{N}$ by $n <_2 m$ if and only if $f(n) \in f(m)$. Then $(\mathbb{N}, <_2) \simeq (\omega + 1, \in)$.
- Consider the map $f\colon \mathbb{N} \to \omega + \omega$ defined by:

$$f(n) = \begin{cases} \omega + \frac{n}{2} & \text{if } n \text{ is even} \\ \frac{n-1}{2} & \text{if } n \text{ is odd.} \end{cases}$$

  Now define the relation $<_3 \subseteq \mathbb{N} \times \mathbb{N}$ by $n <_3 m$ if and only if $f(n) \in f(m)$. Then $(\mathbb{N}, <_3) \simeq (\omega + \omega, \in)$.

In each of the above examples we started with a bijection from $\mathbb{N}$ to an ordinal and then *induced* an order on $\mathbb{N}$. In general we have the following fact:

Suppose that A and B are sets, $R \subseteq B \times B$ is a well order on B and $f\colon A \to B$ is a bijection. Then the *induced order* $<^* \subseteq A \times A$, defined by

$$x <^* y \text{ if and only if } f(x) R f(y),$$

is a well order on A.

Definition 4.1.3 An infinite ordinal α is said to be *countable* if and only if there is a bijection between α and ω.

More generally, we often use the following definition which also applies to finite sets:

Definition 4.1.4 A set A is *countable* iff it is empty or there is a surjection from ω to A. Equivalently, there is an injection from A to ω.

4.2 More Countable Sets

A useful theorem when showing that a set is bijective with another set is:

Theorem 4.2.1 *[Cantor–Schröder–Bernstein] Suppose that there is an injection from A to B and an injection from B to A. Then there is a bijection between A and B.*[2]

Proof (by Gyula König [58]) Let $f: A \to B$ be an injection, and let $g: B \to A$ be an injection. For each $a \in A$ consider a two-sided sequence

$$\dots g^{-1}(a),\ a,\ f(a),\ g(f(a)),\ f(g(f(a)),\dots$$

This sequence can have finitely many or infinitely many elements before a. Let us call each such sequence an *orbit*.

Note that the orbits form a disjoint partition of the elements of $A \cup B$, so it suffices to show that within each orbit there is a bijection between the set of all elements of A that appear on the orbit and the set of all elements of B that appear on the same orbit.

Let us consider an arbitrary orbit $\dots g^{-1}(a),\ a,\ f(a),\ g(f(a)),\ f(g(f(a)),\dots$ There are three cases:

Case 1: The sequence has infinitely many elements before a. Then f is a bijection on this orbit (and so is g).

Case 2: Not Case 1 and the sequence starts with an element of A. Then f is a bijection on this orbit.

Case 3: Not Case 1 and the sequence starts with an element of B. Then g is a bijection on this orbit. $\bigstar_{4.2.1}$

This theorem is really mostly quoted as the Schröder–Bernstein theorem. Cantor proved it first, but his proof used the Axiom of Choice. His student Bernstein found a constructive proof. Schröder found, later and independently, an alternative proof which contained a mistake. The proof we gave also uses the Axiom of Choice, but we like the proof for its simplicity.

4.3 Uncountable Ordinals

So far all the infinite ordinals we have seen have been countable, but one would be wrong if one would have thought that this was the case for all of them. The following is probably Cantor's most celebrated theorem. Along with his proof

[2] We give a history of this theorem after the proof. For a more detailed history see Lorenz Halbeisen's book [48], which also explains why we should not credit the theorem the way we did. We have succumbed to the tradition.

that the set of algebraic numbers is countable, the theorem shows that there are uncountably many transcendental numbers. At the time Cantor wrote his paper [11], basically only three transcendental numbers were known.

Theorem 4.3.1 (Cantor's Diagonal Theorem) *There is no bijection between* $\mathbb{R}$ *and* $\mathbb{N}$.

Proof It suffices to show that $[0, 1)$ is not countable, as a subset of a countable set must be countable. For every $x \in [0, 1)$ choose a decimal expansion starting with 0. and not finishing with an infinite tail of 9s (i.e. rather than $0.399999999999\ldots$ we write 0.4).

Suppose for contradiction that $[0, 1)$ is countable. Therefore we can enumerate it as $\{r_n : n < \omega\}$. We define a real number r: its decimal expansion starts with 0. and the n-th digit after . is chosen to be a digit different than the n-th digit after 0. in r_n, and not equal to 9. Then $r \in [0, 1)$ but $r \neq r_n$ for any n – a contradiction. $\bigstar_{4.3.1}$

Now we have the following corollary.

Corollary 4.3.2 *There exists an uncountable ordinal.*

Proof Let $<^* \subseteq \mathbb{R} \times \mathbb{R}$ be a well order of $\mathbb{R}$. One exists by the Axiom of Choice. Let α be the unique ordinal such that $(\alpha, \in) \simeq (\mathbb{R}, <^*)$ and $f : \alpha \to \mathbb{R}$ be the corresponding order isomorphism. If α were countable then we could find a bijection $g : \mathbb{N} \to \alpha$. But then the bijection $f \circ g : \mathbb{N} \to \mathbb{R}$ would contradict Theorem 4.3.1. $\bigstar_{4.3.2}$

By this last corollary and Theorem 4.0.8 (6) the following definition makes sense.

Definition 4.3.3 Let ω_1 be the least uncountable ordinal.

5
Cardinals

Definition 5.0.1 An ordinal α is a *cardinal* if for all $\beta < \alpha$, β is not bijective with α.

Example 5.0.2 • any natural number, ω and ω_1 are cardinals;
• $\omega + 1$, $\omega + \omega$, $\omega_1 + \omega$ are not cardinals.

Definition 5.0.3 The cardinality of a set A is the unique cardinal κ which is bijective with A. It is denoted by $|A|$.

There is a natural order on the cardinals, induced from the ordinals. So, for example, $0 < 1 < 2 < \cdots < \omega < \omega_1$. Notice that the order on the cardinals can also be defined by saying that $\kappa < \lambda$ if there is an injection from κ to λ but not the other way around.

Next we show that there is no biggest cardinal, i.e. we can produce bigger and bigger cardinals.

Theorem 5.0.4 (Cantor) *There is no set A which admits an injection from $\mathcal{P}(A)$ into A.*

Proof Suppose otherwise and let $f\colon \mathcal{P}(A) \to A$ be an injection, for some set A. Let $B = \{a \in A : a = f(C)$ for some C not containing $a\}$. Consider $f(B) = a$ and ask if $a \in B$. There is no answer to this question, so we have a contradiction. $\bigstar_{5.0.4}$

There is an injection from A to $\mathcal{P}(A)$, namely $a \mapsto \{a\}$. In conclusion, $|A| < |\mathcal{P}(A)|$.

On the basis of this theorem we can, for every cardinal κ, define 2^κ to be the cardinal $|\mathcal{P}(\kappa)|$ and conclude that $\kappa < 2^\kappa$. We can also note that, since $2^\kappa \in \{\alpha \in 2^\kappa + 1 : \alpha > \kappa$ is a cardinal$\}$, this set of ordinals is non-empty and hence has the least element. Therefore there is the least cardinal larger than a given cardinal κ, and we denote it by κ^+.

It takes a bit of thinking to understand that the well ordering of the class of ordinals allows us to make definitions by transfinite recursion: these are definitions for which we define an object O_α for every ordinal α. The recursion has the base step $\alpha = 0$, the successor step $\alpha+1$ – assuming O_β has been defined for $\beta \leq \alpha$, and finally the limit step O_δ for δ a limit ordinal > 0. An example of a definition using this method is given by two hierarchies of cardinals:

Definition 5.0.5 1. $\aleph$-hierarchy. By transfinite recursion we define $\aleph_0 = \omega$, $\aleph_{\alpha+1} = \aleph_\alpha{}^+$, and $\aleph_\delta = \bigcup\{\aleph_\alpha : \alpha < \delta\}$ for δ limit > 0.
2. $\beth$-hierarchy. By transfinite recursion we define $\beth_0 = \omega$, $\beth_{\alpha+1} = 2^{\beth_\alpha}$, and $\beth_\delta = \bigcup\{\beth_\alpha : \alpha < \delta\}$ for δ limit > 0.

The proper way to address recursive definitions and the interplay that they enjoy with proofs by induction, is to keep in mind that definitions are by recursion and proofs by induction. Yet, it is a common misnomer in set theory to talk about definitions and proofs in terms of *induction* and we shall be guilty of it too.

Now we know that the infinite countable cardinalities fit here (i.e. $\omega = \aleph_0 = \beth_0$), but how about the cardinality of the reals? It is in fact easy to locate this cardinality in the $\beth$-hierarchy, because of the following theorem:

Theorem 5.0.6 *There is a bijection between* $\mathbb{R}$ *and* $\mathcal{P}(\omega)$.

Proof Cantor showed that every real number $r > 1$ can be written in a unique way as a product of the form

$$r = \prod_{n\in\omega}\left(1 + \frac{1}{q_n}\right)$$

where all $q_n = q_n(r)$'s are positive integers and for all $n \in \omega$ we have $q_{n+1} \geq q_n^2$. Such products are called *Cantor products*.

Let us use this to show that there is an injection from $\mathbb{R}$ to $\mathcal{P}(\omega)$. First note that there is an injection between $\mathbb{R}$ and $(1, \infty)$ (do this as an exercise). For $r > 1$ let $f(r) = \{q_n(r) : n < \omega\}$. The composition of the two functions is an injection.

We now exhibit an injection from $\mathcal{P}(\omega)$ to $\mathbb{R}$. Consider for example the function $g(x) = \sum_{n\in x} 3^{-n}$, where $g(\emptyset) := 0$, which is obviously an injective mapping from $\mathcal{P}(\omega)$ into $\mathbb{R}$ (or more precisely, into the interval $[0, \frac{3}{2}]$).

The conclusion follows from the Cantor-Schröder-Bernstein theorem. $\bigstar_{5.0.6}$

We normally denote $|\mathbb{R}| = \mathfrak{c}$. Therefore the above theorem says that $\mathfrak{c} = \beth_1$. Locating the value of $\mathfrak{c}$ on the $\aleph$-hierarchy is a different matter. Cantor tried to solve this problem for many years, and so did many others. It was the first problem on Hilbert's celebrated list of problems for the twentieth century, formulated at the International Congress of Mathematics in Paris in 1900. It is the

Continuum Hypothesis (CH) $\mathfrak{c} = \aleph_1$. Equivalently, any infinite subset of the reals is bijective either with $\mathbb{N}$ or with $\mathbb{R}$.

The complete picture of the relation between CH and ZFC was formed only in 1963, when it was shown by Paul Cohen that CH is independent from ZFC. To understand what this means we shall have to go back to some logical issues. Before that, let us also state a generalised version of CH, the so called GCH.

Generalised Continuum Hypothesis (GCH) For every infinite cardinal κ we have $2^\kappa = \kappa^+$. Equivalently, for every ordinal α we have $\aleph(\alpha) = \beth(\alpha)$.

6
Models and Independence

Let $\mathcal{L}$ be a language of first order logic. A *theory* T is a set of $\mathcal{L}$-sentences. T is *consistent* if there is no contradiction ensuing that can be deduced from T using the following notion $\vdash$:

Definition 6.0.1 $T \vdash \varphi$ for a sentence φ if there is a finite list of the form $\varphi_0, \ldots, \varphi_n$ such that $\varphi_n = \varphi$ while each φ_i for $i < n$ is either a sentence from T, or obtained from $\varphi_0, \ldots, \varphi_{i-1}$ as an implication $\bigwedge_{j<i} \varphi_j \implies \varphi_i$.

First order logic is *complete*, which means that for any sentence φ and theory T, $T \vdash \varphi$ iff for any model M of T, $M \models \varphi$. This is written as $T \models \varphi$. The equivalence $T \vdash \varphi \iff T \models \varphi$ is the content of Kurt Gödel's Completeness Theorem [44] (this is *not* Gödel's *In*completeness Theorem [45]).

In some literature 'a theory' stands for 'a consistent theory', but things are usually clear from the context.

Let us ask the question: is ZFC consistent? Because of the Completeness Theorem, this is equivalent to asking if ZFC has a model. We must be careful as to how we ask this question, because of the Incompleteness Theorems, again due to Gödel. They both come from the revolutionary work in the paper [45].

In order to formulate these theorems we have to recall the theory of Peano Arithmetic (PA). It is basically the first order theory of arithmetic on natural numbers. Here are its axioms, in the language $\mathcal{L} = \{S, 0\}$ where 0 is a constant symbol and S is a unary relation symbol:

1. There is no n such that $S(n) = 0$.
2. S is injective.
3. For any φ, if $\varphi(0)$ holds and for every n, $\varphi(n) \implies \varphi(S(n))$, then $\varphi(n)$ holds for all n.

Now we say what it means for a theory to be *effectively* or *recursively* generated. For those who know this concept, it simply means that the set of the axioms of the theory is *recursively enumerable*. To explain what this means, suppose that we are working with a theory T given with a fixed set A of axioms in some language $\mathcal{L}$ in which we can produce only countably many sentences of first order logic (for example the language of set theory is such a language). Let us collect all these sentences in some set S. T is effectively generated if there is an algorithm which will let us choose one by one the elements of S and decide if the chosen element is a member of A or not, and then proceed to the next element of S. We can imagine that there is a computer programme that, if let to work for possibly infinitely many steps, can enumerate all the axioms of the theory without listing any statements that are not the axioms. ZFC is an effectively generated theory.

Here are Gödel's two Incompleteness Theorems (the second one in fact implies the first):

Theorem 6.0.2 *Suppose that T is an effectively generated consistent first order theory which implies the axioms of* PA. *Then there is a sentence φ in the language of T such that T does not prove φ and it also does not prove its negation.*

Note that the above means that such a sentence φ is *independent* of T. Applying this theorem to ZFC, we conclude that if ZFC is consistent, then there is a sentence in the language of set theory which is independent of ZFC.

We may take this as a proof that ZFC is not a good enough theory and that we should strengthen it, or change it. As far as strengthening it goes, it is not going to work out, since the theorem says that if we strengthen ZFC in any reasonable way, we shall still have a sentence which is independent. The theorem also means that any other reasonable axiomatisation of set theory, since it will inevitably imply PA, will have the same problem as ZFC.

The proof of Theorem 6.0.2 is quite complex, but a major idea is a clever use of Cantor's diagonal argument or the so called Liar's paradox. The proof is existential, so we cannot use it to read off the sentence which is independent of the theory. It is therefore absolutely amazing that it turned out for ZFC that a nicely formulated statement CH is independent of ZFC. We shall endeavour to prove this. In order to commence, we shall have to also understand Gödel's Second Incompleteness Theorem.

Theorem 6.0.3 *Suppose that T is an effectively generated first order theory which implies the axioms of* PA, *and which proves that T is consistent. Then T is inconsistent.*

So now suppose that we want to prove the consistency of ZFC. Since we take the line that ZFC axiomatises mathematics (i.e. that all proofs follow from

ZFC), this proof will have to be carried out within ZFC. Applying Gödel's theorem, if we succeed, then we obtain that ZFC is inconsistent!

Therefore the consistency of ZFC cannot be proved in ZFC. It can be proved in some stronger theories, and quite easily, but not within ZFC itself. So we certainly also cannot hope to prove that ZFC+CH is consistent or that ZFC + ¬CH is consistent.

What we can hope for are *relative consistency* results. This means that, using the notation Con(T) to state that a theory is consistent, we shall prove Con(ZFC) $\implies$ Con(ZFC + CH) and Con(ZFC) $\implies$ Con(ZFC + ¬CH).

7
Some Class Models of ZFC

Even though while arguing in ZFC we cannot produce a set model of ZFC, there is nothing to prevent us from producing a model which is a proper class.

7.1 The Cumulative Hierarchy V

By induction on α an ordinal we define

- $V_0 = \emptyset$,
- $V_{\alpha+1} = \mathcal{P}(V_\alpha)$ for any α,
- $V_\delta = \bigcup_{\alpha<\delta} V_\alpha$ for $\delta > 0$ a limit ordinal.

Let $\mathbf{V} = \bigcup_{\alpha\in\mathrm{Ord}} V_\alpha$, where Ord stands for the class of all ordinals.

Theorem 7.1.1 *$\mathbf{V}$ is exactly the class of all sets.*

Proof In one direction, we prove that any element A of $\mathbf{V}$ is a set. This can be done by transfinite induction on α, which is the first ordinal such that $A \in V_{\alpha+1}$ (call such an ordinal the rank $\mathrm{r}(A)$). Notice that the first ordinal β such that $A \in V_\beta$ is never a limit ordinal, so this definition of rank makes sense.

In the other direction suppose that there is a set A which is not in $\mathbf{V}$. Clearly, such a set is non-empty, as $\emptyset \in \mathcal{P}(\emptyset) = V_1$. If all elements of A are in $\mathbf{V}$, then $\{\mathrm{r}(a) : a \in A\}$ is a set of ordinals and, hence, there is a limit ordinal δ larger than all the elements of this set. Hence all elements of A are in V_δ, and so $A \subseteq V_\delta$, and therefore $A \in \mathcal{P}(V_\delta) = V_{\delta+1}$, in contradiction with the choice of A.

We conclude that there must be $a \in A$ which is not in $\mathbf{V}$. Continuing in this manner we obtain an infinite sequence in A of the form $a_0, a_1, a_2, \ldots$ of sets such that $a_{n+1} \in a_n$ (and none of these sets are in $\mathbf{V}$). The existence of such a sequence contradicts the Axiom of Foundation. $\bigstar_{7.1.1}$

Taking a look at the proof of Theorem 7.1.1, we observe that for every set a there is indeed a first ordinal β such that $a \in V_\beta$ and that the first such β is a successor ordinal of the form $\alpha + 1$. As in that proof, we define α to be the *rank* of a and denote it by $\mathrm{r}(a)$.

It is interesting to calculate ranks of some familiar sets. For example, $\mathrm{r}(\emptyset) = 0$ as $\emptyset \in V_1$, $\mathrm{r}(\omega) = \omega$ and identifying $\mathbb{R}$ with $\mathcal{P}(\omega)$ we have that $\mathcal{P}(\omega) = V_{\omega+1}$, so $\mathcal{P}(\omega) \in V_{\omega+2}$, so $\mathrm{r}(\mathcal{P}(\omega)) = \omega + 1$.

Some further properties of the sets V_α are:

- Each V_α is transitive.
- If $\alpha \in \beta$, then $V_\alpha \subsetneq V_\beta$.
- $\alpha \subsetneq V_\alpha$ and $\alpha \in V_{\alpha+1}$.
- If $a \in b$ then $\mathrm{r}(a) < \mathrm{r}(b)$, in fact $\mathrm{r}(b) = \sup\{\mathrm{r}(a) : a \in b\} + 1$, for any b.

These properties can be proved by induction. The next fact is that **V** satisfies all axioms of ZFC.[1] Let us check some of the axioms.

Union: Suppose that $\mathcal{F}$ is a family of sets (in **V**). We know that $\bigcup \mathcal{F}$ is a set so by Theorem 7.1.1 we have that $\bigcup \mathcal{F}$ is in **V**.

Pairing, *Power*, *Infinity*, *Comprehension*, *Replacement*: Similar.

Choice: Let a be a set in **V**, then it is an element of some V_α, so in particular a subset of V_α. Hence to show that every set can be well ordered it suffices to show that every V_α can be well ordered by a relation in **V**, and this is true because we are assuming that choice holds and have proved that every relation (since it is a set) is in **V**.

These points show us that we in fact trivially satisfy all the axioms of ZFC in **V** if we assume these axioms as 'allowed'. But the same argument shows that basically any axiom we would add to ZFC and assume it as 'allowed' in our reasoning, would be satisfied in **V**. This justifies that we do not take the existence of **V** as the proof that ZFC is consistent.

More seriously, under certain assumptions it is possible to build a set model of ZFC out of **V**. To develop this we first need a detour back to the properties of cardinals.

7.2 Regularity and Singularity

Let $(L, \leq)$ be any linear order. A subset A of L is said to be *cofinal* if for all $l \in L$ there is $a \in A$ such that $l \leq a$.

[1] This is not to say that we found a model of ZFC, as **V** is not a set, it is a proper class.

For example, every linearly ordered set is cofinal in itself. The set of natural numbers is cofinal in $\mathbb{R}$ and the singleton $\{\alpha\}$ is cofinal in the ordinal $\alpha + 1$.

For an infinite cardinal number κ we say that it is *regular* if for every cofinal subset A of κ we have $|A| = \kappa$. If κ is not regular we say that it is *singular*.

Therefore $\aleph_0$ is regular, and $\aleph_\omega$ is singular (as exemplified by $\{\aleph_n : n < \omega\}$ which is cofinal in $\aleph_\omega$ but has size only $\aleph_0 < \aleph_\omega$). The following theorem of Felix Hausdorff from [49] implies that $\aleph_1, \aleph_2 \ldots$ are all regular:

Definition 7.2.1 A cardinal λ is a *successor cardinal* if there is some cardinal κ such that $\lambda = \kappa^+$. Otherwise, λ is said to be a *limit cardinal.*

Theorem 7.2.2 (Hausdorff) *Every successor cardinal is regular.*

Proof Suppose that $\lambda = \kappa^+$ and that $A \subseteq \lambda$ is a subset of λ, with $|A| \le \kappa$. Let $\alpha = \bigcup A$. Every element of A is an ordinal $< \lambda$, so of size $\le \kappa$. Hence α is the union of $\le \kappa$ many sets of size $\le \kappa$ and so it has size $\le \kappa$ (check!). Hence $\alpha < \lambda$ and therefore A cannot be cofinal in λ, as shown by $\alpha + 1$. ★$_{7.2.2}$

The limit cardinals $> \aleph_0$ that are easy to construct, such as $\aleph_\omega$, are easily seen to be singular. However we may still like to ask the following question:

Question 7.2.3 Is there an uncountable regular limit cardinal?

Definition 7.2.4 An uncountable regular limit cardinal is called *weakly inaccessible*. A *strongly inaccessible* cardinal κ is one which is weakly inaccessible and in addition satisfies that for all $\lambda < \kappa$ we have $2^\lambda < \kappa$.

Theorem 7.2.5 *Neither the existence of a strongly inaccessible nor the existence of a weakly inaccessible cardinal can be proved in ZFC.*

Proof (outline) Suppose that κ is strongly inaccessible. Then we can prove that V_κ is a model of ZFC. Since the existence of such a model cannot be proved in ZFC, so cannot the existence of a strongly inaccessible cardinal.

However, by the technique of forcing, we can pass from a model of ZFC with a weakly inaccessible cardinal to one in which there is a strongly inaccessible one, so the existence of a weakly inaccessible cardinal cannot be proved in ZFC either. ★$_{7.2.5}$

Cardinal properties such that the existence of a cardinal satisfying the property cannot be proved in ZFC are called *large cardinal properties* and a cardinal satisfying such a property is called a *large cardinal*. The standard reference for the theory of large cardinals is Akihiro Kanamori's book [55], where one can see the surprising connections between these cardinals and properties of the set-theoretic universe.

7.3 The Constructible Universe L

Our construction of **V** allowed us to easily show that the axioms of ZFC are satisfied in **V**, if we assume them as the basis of our arguments. However, there was nothing in our construction to let us determine the size of the power set of any given set in **V**. The constructible universe of Gödel [46] is a refinement of the cumulative hierarchy in which we can compute the values of the power set function, in the sense that GCH is true in **L**.

Definition 7.3.1 (outline) Suppose that $(M, \in)$ is a model of the language $\mathcal{L}$ of set theory. A set $A \subseteq M$ is *definable* if there is a formula φ of $\mathcal{L}$ and parameters a_i $(i < n)$ in M such that $A = \{a \in M : \varphi[a; a_0, \ldots, a_{n-1}]\}$ holds.

Let $\mathrm{Def}(M) = \{A : A \text{ is definable over } M\}$.

The heart of this definition is that while we have no idea of how many subsets of a given set M there are, in terms of the $\aleph$-hierarchy, we can see that there are not too many definable ones. Namely, we easily have that if M is infinite, for $M^{<\omega} = \bigcup_{n<\omega} M^n$, $|M^{<\omega}| = |M|$. There are only countably many possible φ, so there can be at most $|M|$ definable subsets of M.

As examples of definable sets, note e.g. that any interval (a, b) is definable in $\mathbb{R}$ and that $\emptyset$ is definable in any model.

In fact, Definition 7.3.1 is only approximate because in order to write it properly we should discuss the idea of *relativisation*, see the discussion below. The good news is that there exists a formal definition which does exactly what Definition 7.3.1 pretends to do; one can see the details of this in [V§1 [63]]. We shall carry on.

Definition 7.3.2 By induction on α we define

- $L_0 = \emptyset$,
- $L_{\alpha+1} = \mathrm{Def}(L_\alpha)$,
- $L_\delta = \bigcup_{\alpha<\delta} L_\alpha$ for δ limit > 0.

We let $\mathbf{L} = \bigcup_{\alpha\in\mathrm{Ord}} L_\alpha$.

Remark 7.3.3 1. For any $(M, \in)$, $\emptyset, M \in \mathrm{Def}(M)$ and $M \subseteq \mathrm{Def}(M) \subseteq \mathcal{P}(M)$. Each finite set in $\mathcal{P}(M)$ is in $\mathrm{Def}(M)$.

2. The following are the basic properties of **L**:

   - Each L_α is a transitive set, **L** is a transitive class,
   - $\alpha < \beta \implies L_\alpha \subseteq L_\beta$.

Proof (1) We have $M = \{x \in M : x = x\} \in \mathrm{Def}(M)$, $\emptyset = \{x \in M : x \neq x\} \in$

Def(M) and for every $F = \{a_0, \ldots, a_{n-1}\} \subseteq M$, $F = \{x \in M : x = a_0 \vee \cdots \vee x = a_{n-1}\}$.

(2) By induction on α. ★7.3.3

To discuss further properties of **L** we need to discuss the notion of relativisation. Let us first introduce a notation:

Notation 7.3.4 We write $(\exists x \in M)\varphi(x)$ as a shorthand for $(\exists x)(\varphi(x) \wedge x \in M)$ and similarly we use the shorthand $(\forall x \in M)\varphi(x)$.[2] Such quantifiers are called *bounded quantifiers*.

Definition 7.3.5 Suppose that $(M, \in)$ is a model of the language $\mathcal{L}$ of set theory and φ a formula of $\mathcal{L}$. We define the *relativisation* φ^M of φ to M by induction on the complexity of φ:

- If φ is $x = y$ or $x \in y$, then $\varphi^M = \varphi$.
- $(\varphi \wedge \psi)^M = (\varphi^M \wedge \psi^M)$, and similarly for $\vee$ and $\neg$, as well as $\Longrightarrow$ and $\Longleftrightarrow$.
- If φ is $(\exists x)\psi(x)$ then φ^M is $(\exists x \in M)\psi^M(x)$, and similarly for $\forall$.

For example, the relativisation of the formula $(\exists x)(x^2 = -1)$ to $\mathbb{C}$ is $(\exists x \in \mathbb{C})(x^2 = -1)$, which is true, while its relativisation to $\mathbb{R}$ is $(\exists x \in \mathbb{R})(x^2 = -1)$, which is not true. In general note that the definition of $\models$ is set up so that $M \models \varphi$ iff φ^M is true.

The example above shows that there can be a large difference between a formula and its relativisation. This gives rise to the following definition

Definition 7.3.6 Suppose that $M \subseteq N$ are two models of the same language. We say that a formula φ of that language is *absolute* between M and N if φ^N holds iff φ^M holds. If a formula is absolute between any M and N then we say that it is absolute.

There is a large class of formulas that are absolute between transitive models of a sufficient amount of ZFC.

Definition 7.3.7 A formula φ is a Δ_0 formula if all its quantifiers are bounded, or if we define this by induction on complexity:

- if a formula has no quantifiers then it is Δ_0;
- conjunction, disjunction, implication, equivalence and negation applied to Δ_0 formulas give a Δ_0 formula;
- if φ is Δ_0 then so are $(\exists x \in y)\varphi$ and $(\forall x \in y)\varphi$.

[2] Note that this is not necessarily the same as $(\forall x)(\varphi(x) \wedge x \in M)$.

Theorem 7.3.8 *If M and N are classes with $M \subseteq N$, M is transitive, and $\varphi(x_0, \ldots, x_{n-1}; \boldsymbol{a})$ is a Δ_0 formula with $\boldsymbol{a} \in M$, then for all $b_0, \ldots, b_{n-1} \in M$ we have*

$$\varphi^M[b_0, \ldots, b_{n-1}; \boldsymbol{a}] \iff \varphi^N[b_0, \ldots, b_{n-1}; \boldsymbol{a}].$$

Proof By induction on the complexity of φ. If φ is quantifier-free, then this follows because $=$ and $\in$ have the same meaning in M and in N. The connectives step of the induction is clear.

Suppose that $\varphi(x_0, \ldots, x_{n-1}; \boldsymbol{a}) = (\exists u \in a)\psi(u, x_0, \cdots x_{n-1}; \boldsymbol{c})$ for some $a \in M$. If $\varphi^M[b_0, \ldots, b_{n-1}; \boldsymbol{a}]$ holds then by the definition of the relativisation, $(\exists u \in a)^M \psi^M[u, b_0, \ldots, b_{n-1}; \boldsymbol{c}]$ holds for some $\boldsymbol{c}$ in M. Hence there is $u \in M$ such that $u \in a$ and $\psi^M[u, b_0, \ldots, b_{n-1}; \boldsymbol{c}]$ holds. Since $M \subseteq N$, $u \in N$ and by the inductive hypothesis $\psi^N[u, b_0, \ldots, b_{n-1}; \boldsymbol{c}]$ holds. Therefore $\varphi^N[b_0, \ldots, b_{n-1}; \boldsymbol{a}]$ holds. On the other hand, suppose $\varphi^N[b_0, \ldots, b_{n-1}; \boldsymbol{a}]$ holds, so for some $u \in a$ satisfying $u \in N$, $\psi^N[u, b_0, \ldots, b_{n-1}; \boldsymbol{c}]$ holds. Since M is transitive, $a \in M$ and $u \in a$ implies that $u \in M$, so by the inductive hypothesis $\psi^M[u, b_0, \ldots, b_{n-1}; \boldsymbol{c}]$ holds. Therefore $\varphi^M[b_0, \ldots, b_{n-1}; \boldsymbol{a}]$ holds.

The step for $\forall u \in a$ is similar. $\bigstar_{7.3.8}$

Many expressions in set theory can be given by a Δ_0 formula; here is a sample:

Theorem 7.3.9 *The following are expressible by a Δ_0 formula:*

- $z = \{x, y\}$, $x = (u, v)$, $x = \emptyset$, $x \subseteq y$, *'x is transitive', 'x is an ordinal', 'x is a limit ordinal', 'x is a natural number',* $x = \omega$*;*
- *the negations of any of the above statements;*
- $Z = X \times Y$, *similarly for* $\setminus$, $\bigcup$, $\bigcap$, $Z = \mathrm{dom}(f)$, $Z = \mathrm{ran}(f)$*;*
- *'R is a relation', 'f is a function',* $y = f(x)$, $g = f \restriction X$.

Proof Most of these do not need much comment, but we shall comment on the statement 'x is an ordinal'. Note that in general $\mathcal{P}(x)$ is not described by a Δ_0 formula, so the usual definition of an ordinal is not Δ_0. It is easy to see that 'x is linearly ordered by $\in$' is expressible by a Δ_0 formula as is 'x is transitive'. We claim that any transitive set linearly ordered by $\in$ and satisfying $(\forall u \in x)(\forall v \in x)(\forall w \in x)(u \in v \in w \implies u \in w)$ must be an ordinal. So suppose x is such a set yet $\emptyset \neq y \subseteq x$ has no least element. Then we can construct an infinite $\in$-decreasing sequence of elements of y, which contradicts the Axiom of Foundation. $\bigstar_{7.3.9}$

Important examples of statements which are not absolute between transitive models and hence cannot be given by a Δ_0 formula are:

- $y = \mathcal{P}(x)$;
- 'x is a cardinal', 'x is a regular cardinal';
- $x = \omega_1$;
- 'x is bijective with y'.

7.4 Reflection Principle

Definition 7.4.1 Suppose that C is a class, $\varphi(x_0, \ldots, x_{n-1}; \boldsymbol{a})$ is a formula, $\boldsymbol{a} \in C$. The set $M \subseteq C$ is said to *C-reflect* φ if for every $b_0, \ldots, b_{n-1} \in M$

$$\varphi^M[b_0, \ldots, b_{n-1}; \boldsymbol{a}] \iff \varphi^C[b_0, \ldots, b_{n-1}; \boldsymbol{a}].$$

We shall be especially interested in the case when $C = \mathbf{L}$. In the following we use the expression 'reflects' for '$\mathbf{L}$-reflects'. We obtain the following important property:

Theorem 7.4.2 (Reflection Principle) *First assume only the axioms of* ZF. *Let $\varphi(x_0, \ldots, x_{n-1}; \boldsymbol{a})$ be a formula of the language of set theory and $M_0 \subseteq \mathbf{L}$, while $\boldsymbol{a} \in \mathbf{L}$. Then there is a limit ordinal α such that $M_0 \subseteq L_\alpha$ and L_α $\mathbf{L}$-reflects $\varphi(x_0, \ldots, x_{n-1}; \boldsymbol{a})$.*

If we further assume that $\mathbf{L}$ is well ordered, then we can find $M \subseteq \mathbf{L}$ with $M_0 \subseteq M$ and $|M| \leq |M_0| + \aleph_0$, such that M reflects φ.

Similar statements are also true about $\mathbf{V}$.

To prove Theorem 7.4.2, we need the following definition:

Definition 7.4.3 For $a \in \mathbf{L}$ we let $\rho_{\mathbf{L}}(a) \stackrel{\text{def}}{=} \min\{\alpha : a \in L_{\alpha+1}\}$. We call this function the $\mathbf{L}$-rank.

As in the case of $\mathbf{V}$, this rank is well defined.

Proof The proof goes by cases involving the complexity of φ, or by induction on its complexity. The main case is $\varphi = (\exists x)\psi$, so we shall do that case.

Suppose that $b_0, \ldots, b_{n-1} \in M_0$, and $(\exists x \in \mathbf{L})(\varphi^{\mathbf{L}}(x, b_0, \ldots, b_{n-1}; \boldsymbol{a}))$. Hence $\{\alpha : (\exists x)(\rho_{\mathbf{L}}(x) = \alpha \wedge \varphi^{\mathbf{L}}(x, b_0, \ldots, b_{n-1}; \boldsymbol{a}))\} \neq \emptyset$ and has a minimal element, say $\alpha = \alpha_{b_0,\ldots,b_{n-1}}$. Let

$$\delta = \sup(\{\alpha_{b_0,\ldots,b_{n-1}} + 1 : b_0, \ldots, b_{n-1} \in M_0\} \cup \{\rho_{\mathbf{L}}(a) + 1 : a \in M_0\}).$$

Then L_δ is as required.

If we know that $\mathbf{L}$ is well ordered by a relation $<^*$, then we can for every relevant $b_0, \ldots, b_{n-1} \in M_0$ find the $<^*$-least element of $\mathbf{L}$ with $\varphi^{\mathbf{L}}(x, b_0, \ldots, b_{n-1}; \boldsymbol{a})$, and let M be M_0 union the set of all so chosen x. It is easy to see that $|M| \leq |M_0| + \aleph_0$.

We observe that the same proof works for $\mathbf{V}$. $\bigstar_{7.4.2}$

Using reflection arguments we can prove the following important theorem of Montague from [81]. It applies to both ZF and to ZFC; one should read ZF(C) as either ZF or ZFC.

Corollary 7.4.4 *If* ZF(C) *is consistent then it is not finitely axiomatisable.*

Proof Suppose that $\varphi_0, \ldots, \varphi_n$ are finitely many statements axiomatising ZF or ZFC, as required. Let φ be their conjunction, so φ is a single axiom for ZF(C). Applying Theorem 7.4.2 to $M_0 = \emptyset$ and $\mathbf{V}$ we obtain a set M which models φ (as reflecting it from $\mathbf{V}$ in this instance means modelling it!). Hence we have a set model for ZFC, contradicting Gödel's second Incompleteness Theorem. $\bigstar_{7.4.4}$

7.5 ZFC in L

Theorem 7.5.1 *Assume the axioms of* ZF. *Then* $\mathbf{L}$ *is a model of* ZFC.

Proof (outline) We shall give detailed proof of most of the axioms of ZF and a sketch of the proof of the Axiom of Choice in $\mathbf{L}$.

Foundation: Clear, as the meaning of $\in$ in $\mathbf{L}$ and $\mathbf{V}$ is the same.

Extensionality: Suppose that $x, y \in \mathbf{L}$ and $(\forall z)(z \in x \iff z \in y)$ holds in $\mathbf{L}$. This can be rewritten as a Δ_0 formula, so applying the Δ_0-absoluteness to the transitive classes $\mathbf{L} \subseteq \mathbf{V}$, we obtain that $(\forall z)(z \in x \iff z \in y)$ holds in $\mathbf{V}$. Hence $x = y$ holds in $\mathbf{V}$ and also in $\mathbf{L}$.

Pairing: Suppose that $x, y \in \mathbf{L}$, say $x, y \in L_\alpha$. Then $\{x, y\} = \{a \in L_\alpha : a = x \vee a = y\}$ is a definable subset of L_α, and hence belongs to $L_{\alpha+1} \subseteq \mathbf{L}$.

Union: Similar.

Infinity: Note that $\omega \in \mathbf{L}$.

Power set: Let $a \in \mathbf{L}$. It is *not* necessarily true that $\mathcal{P}(a) \in \mathbf{L}$. However, let $b = \mathcal{P}(a) \cap \mathbf{L}$. Therefore there is α such that $b \subseteq L_\alpha$, $a \in L_\alpha$ and moreover then $b = \{x \in L_\alpha : x \subseteq a\}$. Hence $b \in L_{\alpha+1} \subseteq \mathbf{L}$. It is easy to check that $b = \mathcal{P}(a)^{\mathbf{L}}$.

Comprehension Scheme: Let X be a set in $\mathbf{L}$, say $X \in L_\alpha$ and φ a formula. Then $\{x \in X : \varphi(x)\} = \{x \in L_\alpha : x \in X \wedge \varphi(x)\}$ (by the transitivity of L_α), and hence this is in $L_{\alpha+1}$.

Replacement Scheme: Suppose that $X \in \mathbf{L}$ and we have that for some formula φ, for every $x \in X$ there is exactly one $y \in \mathbf{L}$ satisfying $\varphi(x, y)$. By the Reflection Principle, there is a limit ordinal α such that for every $x \in X$ there is exactly one $y \in L_\alpha$ satisfying $\varphi(x, y)$. We may, by increasing α if necessary, assume that $X \in L_\alpha$, so then $\{y : (\exists x \in X)\varphi(x, y)\}$ is in $L_{\alpha+1}$.

This shows that ZF is satisfied. The fact that choice is satisfied follows from the fact that $\mathbf{L} \models$ '$\mathbf{V} = \mathbf{L}$', see below. $\bigstar_{7.5.1}$

We are now going to proceed to give some extra facts about **L** without proving them.

7.6 Facts About L

The *Axiom of Constructibility* states that all sets are constructible, or simply stated that $\mathbf{V} = \mathbf{L}$.

One can rush to state that **L** satisfies the Axiom of Constructibility, as clearly all sets in **L** are constructible. However, recalling the relativisations, to show that **L** satisfies the Axiom of Constructibility, we need to prove $(\mathbf{V} = \mathbf{L})^{\mathbf{L}}$, which translates as 'in **L** we can demonstrate that all sets in **L** are constructible in **L**', i.e. the constructibility has to be witnessed within **L**. This is not trivial, but it is true:

Theorem 7.6.1 (Gödel) **L** *satisfies the Axiom of Constructibility.*

We shall give an outline of the proof. For it we need to extend our understanding of formulas.

Definition 7.6.2 (Lévy hierarchy of formulas) By induction on $n < \omega$ we define classes Σ_n, Π_n, Δ_n of formulas.

$\underline{n = 0}$ $\Sigma_0 = \Pi_0 = \Delta_0$.

$\underline{n + 1}$ A formula is Σ_{n+1} if it is of the form $(\exists x)\,\varphi$ for some Π_n formula φ. A formula is Π_{n+1} if it is of the form $(\forall x)\,\varphi$ for some Σ_n formula φ. A formula is Δ_{n+1} if it is both Σ_{n+1} and Π_{n+1}.

A relation is of complexity Σ_n, Π_n, Δ_n if it can be expressed using a Σ_n, Π_n, Δ_n formula (respectively).

Therefore the Lévy hierarchy exhausts all formulas of the language of set theory. The importance for us is the following:

Lemma 7.6.3 *Suppose that $M \subseteq N$ are classes and M is transitive. Then*

- *for any Σ_1 formula φ, for all $a \in M$, if $\varphi^M(a)$ holds then $\varphi^N(a)$ holds,*
- *for any Π_1 formula φ, for all $a \in M$, if $\varphi^N(a)$ holds then $\varphi^M(a)$ holds,*
- *for any Δ_1 formula φ, for all $a \in M$, $\varphi^N(a)$ holds iff $\varphi^M(a)$ holds.*

Proof of the lemma Suppose that φ is Σ_1, hence $\varphi(y) = (\exists x)\,\psi(x, y)$ for some Δ_0 formula $\psi(x, y)$. Hence for $a \in M$, $\varphi^M(a) = (\exists x \in M)\psi^M(x, a)$, which by transitivity of M implies that $(\exists x \in M)\psi^N(x, a)$, but then since $M \subseteq N$, we have $(\exists x \in N)\psi^N(x, a)$ and therefore $\varphi^N(a)$. This proves the first item of the lemma, the others are similar. $\bigstar_{7.6.3}$

Now we need the following lemma, which we give without a proof (see [51] for a complete proof).

Lemma 7.6.4 (Gödel) *The function* $\alpha \mapsto L_\alpha$ *is* Δ_1.

Definition 7.6.5 An *inner model* of ZF is a transitive class which contains the ordinals and satisfies the axioms of ZF.

For example, **L** is an inner model.

Corollary 7.6.6 *The property 'x is constructible' is absolute for inner models of* ZF.[3]

Proof of the corollary Let M be an inner model. Since M contains the ordinals, we have that for $x \in M$,

$$(x \text{ constructible})^M \iff (\exists \alpha \in M)\, x \in L_\alpha^M \iff (\exists \alpha) x \in L_\alpha \iff x \text{ constructible}.$$

$\bigstar_{7.6.6}$

Outline of the proof of Theorem 7.6.1 Suppose that $x \in \mathbf{L}$. Then by the definition of **L**, 'x is constructible'. By Corollary 7.6.6 $(x \text{ is constructible})^{\mathbf{L}}$ holds. $\bigstar_{7.6.1}$

The above considerations, by the same proof, also lead to the following important fact about **L**:

Theorem 7.6.7 (Gödel) *Suppose that M is an inner model of* ZF. *Then* $\mathbf{L}^M = \mathbf{L}$. *In particular* $\mathbf{L} \subseteq M$, *i.e.* **L** *is the smallest inner model of* ZF.

Important consequences of $\mathbf{V} = \mathbf{L}$ are given by the following:

Theorem 7.6.8 (Gödel) *Suppose that* $\mathbf{V} = \mathbf{L}$. *Then:*

- *the Axiom of Choice holds and*
- GCH *holds.*

[3] i.e. the property is absolute between such a model and **V**.

Proof (outline) For the Axiom of Choice, we can construct a *global* well ordering of **L**. So not only that every set in **L** can be well ordered separately, but the whole class **L** can be well ordered. The proof of this is to construct a well ordering $<_\alpha$ on L_α by induction on α, so that $\alpha < \beta \implies <_\alpha \subseteq <_\beta$.

For GCH, one can show that for every α and every $x \in \mathcal{P}(L_{\omega_\alpha})^{\mathbf{L}}$, we have $x \in L_{\omega_{\alpha+1}}$. Since the size of $L_{\omega_{\alpha+1}}$ can easily be shown (by induction on α) to be $\aleph_{\alpha+1}$, this finishes the proof. ★7.6.8

7.7 Mostowski Collapse

A corollary of the arguments presented above is an interesting consistency result.

Theorem 7.7.1 Con(ZF) $\implies$ Con(ZFC+**V** = **L**)*, in particular* Con(ZF) $\implies$ Con(ZFC + GCH).

Seeing this result, it is tempting to think that Gödel discovering **L** would mean that he believed it possible that ZFC implies GCH. In fact, it is just the opposite. He had constructed **L** while hoping he would find a proof that ZFC does not imply GCH. It can be seen from his writing that already in 1949 he had thought of the possibility of the independence of GCH.

The full proof of Theorem 7.7.1 is not completely straightforward. However, there is a weaker version of it which we can prove quite easily. To prove it, we shall introduce the Mostowski collapse, which is a technique developed by Andrzej Mostowski in [83].

Definition 7.7.2 A relation E on a set A is *extensional* if for every $a \neq b$ in A we have $\{c : cEa\} \neq \{c : cEb\}$.

Theorem 7.7.3 (Mostowski collapse) *Suppose that M is a class (or a set) and E is an extensional well founded relation on M. Then there is a unique transitive class (respectively a set) N and a bijection $\pi\colon M \to N$ satisfying $aEb \iff \pi(a) \in \pi(b)$ for $a, b \in M$ (in other words, $(N, \in)$ is isomorphic to (M, E)).*

The set N as in the Mostowski collapsing theorem is called the Mostowski collapse of (M, E).

Proof of the Mostowski collapse Since E is well founded, we can define objects by recursion on E (check this!). Therefore the function $\pi(z) = \{\pi(x) : xEz\}$ is well defined. Because E is extensional, π is injective. Let N be the range of π. Then π is an isomorphism between (M, E) and $(N, \in)$. Suppose that $a \in N$, so there is $z \in M$ such that $\pi(z) = a$. Hence $a = \{\pi(x) : x \in z\}$, so $a \subseteq N$. Therefore N is transitive. $\bigstar_{7.7.3}$

This is the weak version of Theorem 7.7.1 which we shall actually prove.

Theorem 7.7.4 *Suppose that there is a well founded model of ZF. Then there is a model of ZFC+*$\mathbf{V} = \mathbf{L}$.

Proof Suppose that (M, E) is a well founded model of ZF, where E is the interpretation of $\in$. In particular, E is extensional and well founded. Hence there is a unique Mostowski collapse N of M. Then $(N, \in)$ is a model of ZF, as all the axioms are preserved by isomorphism. By the arguments about $\mathbf{L}$ applied within N, N satisfies that $\mathbf{L}^N$ is an inner model of ZFC + $\mathbf{V} = \mathbf{L}$. Therefore $N \models (\mathbf{L}^N \models \text{ZFC}+\mathbf{V} = \mathbf{L})$. We claim that $\mathbf{V} \models (\mathbf{L}^N \models \text{ZFC}+\mathbf{V} = \mathbf{L})$. To check this we need to verify that the notion $\models$ is upwards absolute between N and $\mathbf{V}$. This is where we shall use that N is transitive. So suppose that φ is a first order sentence of set theory, $a \in N$ and $N \models ((a, \in) \models \varphi)$. The proof is by induction on the complexity of φ.

If φ has no quantifiers then the conclusion follows because $\in$ and $=$ are interpreted by their true values, as N is transitive. The case of the connectives is simple. If $\varphi = (\exists x)\,\psi$ then $a \models (\exists b)\psi(b)$. Since N is transitive, $a \subseteq b$, so N satisfies $(\exists b)\psi^a(b)$. Therefore the same is true in $\mathbf{V}$. $\bigstar_{7.7.4}$

Note that not every model (M, E) of ZF can be assumed to be well founded, in spite of the Axiom of Foundation (this seems counter-intuitive, but we remind the reader that the Axiom of Choice is not assumed in the argument).

7.8 Trees and the Fine Structure of L

After Gödel's work, research on **L** lay dormant for many years, to be gloriously resuscitated by Ronald Jensen in the 1970s, discovering that **L** has a lot of hidden structure. This makes it amenable to combinatorial arguments known as *fine structure theory* as much as to an analysis and generalisations of its properties as an *inner model*, which is the start of the field of inner model theory. The founding paper for this subject is Jensen's celebrated [54]. We shall discuss one of the most well known combinatorial results of [54]. It relates to Souslin's problem, which we now describe.

The following is Cantor's characterisation ([128], p. 310) of the ordering of the real numbers $(\mathbb{R}, <)$: any linear order $(L, <)$ which

- has no first or last element,

- is dense, that is if $a < b$ then there is c such that $a < c < b$,
- is complete, in the sense that every non-empty bounded subset has an infimum and a supremum,
- (separability) has a countable dense set,

is isomorphic to $(\mathbb{R}, <)$. Mikhail Souslin in [116] asked the question if the characterisation can be weakened by replacing the separability by the condition known as ccc: every family of disjoint open intervals is countable. A purported example of a linear order $(L, <)$ which satisfies the properties proposed by Souslin but is not isomorphic to $(\mathbb{R}, <)$ is known as a *Souslin line*. The hypothesis that there are no such lines is known as the Souslin Hypothesis (SH).

Jensen [54] proved that one can construct a Souslin line in **L** and we shall describe his proof. In §12.1, we shall see a proof that it is relatively consistent with ZFC that there is a model with no Souslin lines. Therefore SH is independent of ZFC. To show Jensen's proof, we shall first review the work of Nachman Aronszajn and Đuro Kurepa, both described in Kurepa's 1935 thesis at the Sorbonne, [64]. The idea is to replace lines by partial orders known as trees.

Definition 7.8.1 (1) A tree is a partial order $(T, <_T)$ in which for every $x \in T$ the set $\{y \in T : y <_T x\}$ is well ordered. The order type of $\{y \in T : y <_T x\}$ is called the *height* of x and is denoted by $\mathrm{ht}(x)$. A *level* of T is a non-empty set of elements of T of the same height, say β, denoted by $\mathrm{lev}_\beta(T)$. The *height* of T is the first ordinal α such that $\mathrm{lev}_\alpha(T)$ is not defined.
(2) An ω_1-tree is a tree of height ω_1 in which all levels are countable.
(3) A *chain* in a partial order is a linearly ordered subset. An ω_1-tree with no uncountable chains is called an *Aronszajn tree*.
(4) An *antichain* in a partial order is a set of incomparable elements. An Aronszajn tree with no uncountable antichains in called a *Souslin tree*.
(5) An Aronszajn tree which is a countable union of antichains is called *special*.

Another notion often used in the context of trees is that of a *branch*, where a branch is a maximal chain.

By considering the tree of intervals in a line, ordered by reverse inclusion, Kurepa [64] proved that there is a Souslin line iff there is a Souslin tree. Research on SH has since concentrated to constructing a Souslin tree. Clearly, a Souslin tree cannot be special. The following is a theorem of Aronszajn.

Theorem 7.8.2 (Aronszajn) *There is an Aronszajn tree.*

Kurepa in [64] gives Aronszajn's original construction of an Aronszajn tree, which uses convergent series of rational numbers. The construction by Kenneth Kunen in [63] uses the idea of injective embeddings of countable ordinals into ω and finally, another construction of such a tree is due to Stevo Todorčević in [119], who uses his method of minimal walks. All of these constructions give a special Aronszajn tree.

Jensen [54] proved:

Theorem 7.8.3 (Jensen) *There is a Souslin tree in* **L**.

Jensen's proof relies on the following combinatorial principle, known as $\diamond$:

Principle $\diamond$: there is a sequence $\langle A_\alpha : \alpha < \omega_1 \rangle$ such that $A_\alpha \subseteq \alpha$ and for every $A \subseteq \omega_1$, the set

$$\{\alpha < \omega_1 : A_\alpha = A \cap \alpha\}$$

is stationary.

To understand this principle, we need to make a small detour into an important part of combinatorial set theory, which deals with clubs and stationary sets.

Definition 7.8.4 Suppose that α is an ordinal and $C \subseteq \alpha$.
(1) C is said to be *unbounded in* α if $\sup(C) = \alpha$.
(2) C is said to be *closed in* α if C is closed in the order topology induced by the ordinal order of α, namely, for every $\beta < \alpha$ that satisfies $\sup(C \cap \beta) = \beta$, we have $\beta \in C$.
(3) C is a *club* of α if it is closed and unbounded.
(4) $S \subseteq \alpha$ is stationary if $S \cap C \neq \emptyset$ for any club C of α.

Clubs and stationary sets are of particular interest if α is a cardinal of uncountable cofinality, usually denoted by κ. In this case, the club sets form a family that is closed under the intersection of $< \mathrm{cf}(\kappa)$ elements. Suppose now that κ is a regular uncountable cardinal, since that is the case that interests us the most. Then for any sequence $\langle C_\alpha : \alpha < \kappa \rangle$ of clubs of κ, the *diagonal intersection*

$$\Delta_{\alpha<\kappa} C_\alpha = \{\beta : (\forall \alpha < \beta) \beta \in C_\alpha\}$$

is a club. Finally, the most important property of stationary sets has to do with the following notion of regressive functions.

Definition 7.8.5 A function $f : \alpha \to \alpha$ for an ordinal α is *regressive* if for every $\beta < \alpha$ we have $f(\beta) < \beta$.

A close connection between stationary sets and regressive functions was discovered by Géza Fodor in [31], as follows.

Theorem 7.8.6 *[Fodor] Suppose that κ is a regular uncountable cardinal and $S \subseteq \kappa$ is stationary, while $f : S \to \kappa$ is regressive. Then there is a stationary subset T of S such that $f \restriction T$ is constant.*

Now that the reader can appreciate what the statement of $\diamond$ says, let us describe how it is used. Jensen proved that, on the one hand, $\diamond$ holds in **L**, and on the other hand $\diamond$ implies the existence of a Souslin tree. Let us give a similar proof obtained by Alexander Primavesi in his Ph.D. thesis at UEA [89], where he uses a seemingly weaker principle, named Superclub. It is not known if Superclub is actually weaker than $\diamond$.

Principle Superclub : there is a sequence $\langle A_\alpha : \alpha < \omega_1\rangle$ such that for any unbounded $X \subseteq \omega_1$, there is an unbounded $Y \subseteq X$ such that the set

$$\{\alpha \text{ limit } < \omega_1 : A_\alpha = Y \cap \alpha \text{ is an unbounded subset of } \alpha\}$$

is stationary.

Proof (of Theorem 7.8.3) We prove that Superclub implies the existence of a Souslin tree. The proof is very similar to the original proof by Jensen.

Let $\langle A_\alpha : \alpha < \omega_1\rangle$ be a witness to the Superclub. We construct a tree ordering $<_T$ on $\omega_1 \setminus [1, \omega)$ by inductively defining the restriction of $<_T$ to $[\omega\alpha, \omega\alpha + \omega)$, for $1 \leq \alpha < \omega_1$. We shall have that the following inductive hypothesis $(*)$ holds:

- $\mathrm{lev}_\beta(T) = [\omega\beta, \omega\beta + \omega)$ for any $1 \leq \beta < \omega_1$ and
- if $\alpha < \beta < \omega_1$ and $x \in \mathrm{lev}_\alpha(T)$, there is $y \in \mathrm{lev}_\beta(T)$ with $x <_T y$.

We let $\mathrm{lev}_0(T) = \{0\}$ and let $0 <_T \omega + n$ for all $n < \omega$.

If $\alpha = \beta + 1$ for some $\beta > 0$, then for $y = \omega\beta + n \in \mathrm{lev}_\beta(T)$, we let $y <_T \omega\alpha + 2n$ and $y <_T \omega\alpha + (2n + 1)$.

Suppose now that $0 < \alpha < \omega_1$ is a limit ordinal. Let $T_\alpha = \bigcup_{\beta<\alpha} \mathrm{lev}_\beta(T)$. Note that we are in the situation that $\beta < \alpha \implies \omega\beta < \alpha$, hence $\alpha = \omega\alpha$. We shall need the following lemma.

Lemma 7.8.7 *For each $x \in T_\alpha$, there is a chain B_x of T_α which contains x and which intersects every level of T_α. Moreover, if A_α is a maximal antichain in T_α, then B_x can be chosen so that it contains an element of A_α.*

Proof Let $x \in T_\alpha$ be given. To find B_x, we first choose a cofinal strictly increasing sequence $\langle \zeta_n : n < \omega\rangle$ in α such that $\zeta_0 = \mathrm{ht}(x)$. Then, by induction on n we shall choose a $<_T$-increasing sequence $\langle y_n : n < \omega\rangle$ in T_α with $\mathrm{ht}(y_n) \geq$

ζ_n and such that $y_0 = x$. In addition, if we are in the situation that A_α is a maximal antichain in T_α, we shall require that y_1 satisfies $w <_T y_1$ for some element w of A_α. To see how to perform this induction, let us first discuss the case of y_1.

In the case that A_α is a maximal antichain in T_α, we have that x is $<_T$-comparable with some element w of A_α. If $x \geq_T w$ for some such w, then we simply choose y_1 of height ζ_1 with $x <_T y_1$, which is possible by the assumption $(*)$. If not, then there is $w \in A_\alpha$ with $x <_T w$. By $(*)$, there is $y_1 \geq_T w$ of height at least ζ_1, which we choose for our y_1. If it is not the case that A_α is a maximal antichain in T_α, then we simply choose $y_1 >_T x$ of height ζ_1 by using $(*)$.

Further steps of the induction follow the logic of the previous sentence, using the assumption $(*)$ repeatedly.

At the end we set $B_x = \{z \in T_\alpha : (\exists n < \omega)\, x_n \leq_{T_\alpha} z \text{ or } z \leq_{T_\alpha} x_n\}$. ★7.8.7

Having established Lemma 7.8.7, we continue the proof of the theorem. Enumerate T_α as $\langle x_n : n < \omega\rangle$ and for $y \in T_\alpha$ let $y <_T \omega\alpha + n$ iff $y \in B_{x_n}$. It is easy to see that this step of the construction preserves the inductive hypothesis.

At the end, we have defined $T = \bigcup_{\alpha<\omega_1} \mathrm{lev}_\alpha(T)$ and let us now show that it is a Souslin tree. Clearly, T has height ω_1; let us see that it has no uncountable antichains.

Suppose for a contradiction that A is an uncountable antichain in T. Without loss of generality, by extending A if necessary, A is a maximal antichain. We can observe that then there is a club set C of limit ordinals $\alpha > 0$ such that $\alpha = \omega_\alpha$ and such that $A \cap \alpha$ is a maximal antichain in T_α. By the Superclub principle, there is an uncountable set $B \subseteq A$ and a stationary $S \subseteq C$ such that for every $\alpha \in S$ we have $A_\alpha = B \cap \alpha$.

Let us consider $T' = \{x \in T : (\exists y \in B)\, x \leq_T y \vee y \leq_T x\}$. Clearly, T' is a subtree of T and B is a maximal antichain in T'. We can find $\alpha \in S$ such that $B \cap \alpha = A_\alpha$ is a maximal antichain in T_α. But then the construction assures that $B \cap \alpha$ is already a maximal antichain in T', a contradiction.

We have now shown that T has no uncountable antichains. But any uncountable chain in T would lead to an uncountable antichain by the fact that every node has two incompatible extensions at the next level. That is, if

$$\langle a_\alpha : \alpha < \omega_1\rangle$$

were to be an uncountable chain with $\mathrm{ht}(a_\alpha) = \zeta_\alpha$ for some increasing sequence $\langle \zeta_\alpha : \alpha < \omega_1\rangle$ of ordinals, we could choose a b_α of height $\zeta_\alpha + 1$ with $a_\alpha <_T b_\alpha$ yet $b_\alpha \neq a_{\alpha+1}$. Hence, $\langle b_\alpha : \alpha < \omega_1\rangle$ would be an uncountable antichain, a contradiction. ★7.8.3

Superclub is interesting because it is a strengthening of the principle $\clubsuit$, which itself is a weakening of $\diamond$. Principle $\clubsuit$ proved very useful in various constructions after its introduction by Adam Ostaszewski in [86]. In 1975 Istvan Juhasz asked if $\clubsuit$ is sufficient to construct a Souslin tree. That question is still open, in spite of much activity and various partial solutions.

If one is interested in Souslin trees at higher cardinals, for example the $\aleph_2$-Souslin trees, defined analogously to the Souslin tree on $\aleph_1$ (a tree of size $\aleph_2$ with no chains or antichains of size $\aleph_2$), one can still show that they exist in **L**, but the construction becomes a bit more complicated and in order to pass the limit stages of cofinality ω_1, one uses another combinatorial principle discovered by Jensen and present in **L**. That is the square principle which we define now.

Definition 7.8.8 Suppose that κ is an uncountable cardinal. The principle $\square_\kappa$ posits the existence of a sequence $\langle C_\alpha : \alpha \text{ limit } < \kappa^+\rangle$ such that

- each C_α is a club subset of α and $\mathrm{otp}(C_\alpha) < \alpha$,
- if α is a limit point of C_β, then $C_\alpha = C_\beta \cap \alpha$.

The situation with higher Aronszajn trees and Souslin trees in **L** and in some other models is surveyed in [II. 5 [63]], up to the 1980s. In particular, if $\kappa^{<\kappa} = \kappa$, Ernst Specker in [115] has shown that there is a κ^+-Aronszajn tree and moreover it is special, meaning that it is a union of κ antichains. The topic has continued to attract attention of set theorists. As far as the Aronszajn trees go, there are many contemporary papers on the so called *tree property* which states for a cardinal κ that κ has the tree property if every κ-tree has a κ branch. Therefore $\aleph_0$ has the tree property and $\aleph_1$ does not. It is a very striking result of William Mitchell in [77] that if there is Mahlo cardinal[4], then there is a forcing extension in which there are no special ω_2-Aronszajn trees, and in fact this is an equiconsistency result: if there are no special ω_2-Aronszajn trees, then ω_2 of **V** is Mahlo in **L**. One of the hardest open problems in modern set theory was asked by Menachem Magidor in the 1980s: modulo large cardinals, can we find a model in which $\aleph_1$ is the only regular cardinal which does not have the tree property? A recent paper on the subject, which in addition to its own results, summarises the connection between the tree property and several other concepts in set theory and gives a good bibliography, is [19] by James Cummings, Yair Hayut, Magidor, Itay Neeman, Dima Sinapova and Spencer Unger. We return to the tree property in §18.2.

Since every Souslin tree is in particular an Aronszajn tree, the situation with higher Souslin trees is even more complex. Richard Laver and Saharon Shelah

[4] A *Mahlo cardinal* is one which admits a stationary set of inaccessibles below it.

in [67] showed that from the assumption of the existence of a weakly compact cardinal it follows that it is consistent that CH holds and there are no $\aleph_2$-Souslin trees. Note that by Specker's result mentioned above, in that model there is necessarily an $\aleph_2$-Aronszajn tree. It is not known if one can get GCH in a model in which there are no $\aleph_2$-Souslin trees. A huge research programme on higher Souslin trees is led by Assaf Rinot. Many relevant results are described in his paper [91], which deals with diamonds, Souslin trees and various related objects.

8
Forcing

8.1 Logic of Forcing

So far we have seen a way of building inner models of ZFC from the existing models. Forcing is a way of extending a given model of a portion of ZFC to another one. In fact, forcing is technically speaking, only applicable to countable transitive models (we shall discuss why later). The logic behind this process is as follows.

Let us take for example the consistency of the negation of CH. We want to show Con(ZFC) $\implies$ Con(ZFC + $\neg$CH). Hence suppose that this is not true. Therefore ZFC $\vdash$ CH, by the definition of consistency and provability. Therefore, by the definition of $\vdash$, there is a sequence $\varphi_0, \dots, \varphi_n = \text{CH}$ of sentences, each of which is either an axiom of ZFC or obtained from the previous sentences and axioms of ZFC by logical deduction. The number of axioms of ZFC mentioned in any of this is finite, let us say that they form a finite set S. We assume Con(ZFC), so there is a model (N, E) of ZFC. Here we use the fact that we are working in the first order logic and that Gödel's Completeness Theorem applies.[1].

Then by reflection within N, we can find a countable submodel N_0 of N in which all the axioms of S and Extensionality and Foundation are true. By some extra work which we do not explain here, we can find such a model which is also well founded, using that S is a finite set of axioms. By Mostowski collapse we can find a transitive isomorph M of (N_0, E), in particular M is countable transitive and satisfies all the axioms of S.

Now suppose that we know how to force to obtain $\neg$CH. This means we are able to extend our model M to another model $M[G]$ which satisfies the same amount of ZFC that M does (this is provided by the general theory of forcing),

[1] There is no forcing for second order logic, for example.

and satisfies ¬CH. But then all the axioms from S are true in $M[G]$, yet CH is false, a contradiction with the choice of S.

From now on, we shall use the notation ZFC* to stand for a large enough finite fragment of ZFC, determined individually in every forcing argument.[2]

8.2 Combinatorics of Forcing

Definition 8.2.1 1. A *forcing notion* is a partial order $\mathbb{P}$ with the least element, denoted by $\emptyset_{\mathbb{P}}$. Elements of $\mathbb{P}$ are usually called *conditions*.

2. A subset D of $\mathbb{P}$ is said to be *dense* if $(\forall p \in \mathbb{P})(\exists q \in D)(q \geq p)$.
3. A subset F of $\mathbb{P}$ is said to be a *filter* if
   - $(\forall p \in F)(\forall q \leq p)\, q \in F$,
   - $(\forall p, q \in F)(\exists r \in F)\, r \geq p, q$.
4. Two conditions p, q are said to be *incompatible* if there is no condition $r \geq p, q$. This is denoted by $p \perp q$.[3]
5. Suppose that $\mathcal{F}$ is a family of dense subsets of $\mathbb{P}$. A filter F is said to be $\mathcal{F}$*-generic* if $F \cap D \neq \emptyset$ for any $D \in \mathcal{F}$. If $\mathcal{F}$ is the family of all dense subsets of $\mathbb{P}$ contained in some model M then an $\mathcal{F}$-generic filter is said to be $\mathbb{P}$*-generic* over M.

We may think of a forcing construction as a generalised inductive (recursive) construction. In an inductive construction we construct an object by giving its approximations over a linearly ordered set, for example ω. At every time $n < \omega$ we have a partial, n-th information about the object, and these pieces of information agree in that ones corresponding to a larger n agree with what was already said by the smaller ones. The approximations are 'glued together' using some operation, usually the union.

In forcing, we may think of every $p \in \mathbb{P}$ as giving us some information about the desired object we wish to construct. Say we want to construct a subset of ω, which we may identify with its characteristic function $f\colon \omega \to 2$. Then we can define

$$\mathbb{P} = \{s : s\colon F \to 2, F \text{ finite} \subseteq \omega\} \tag{8.1}$$

and we order $\mathbb{P}$ by letting $p \leq q$ iff $p \subseteq q$.[4] Note that this is really a forcing notion, with the least element $\emptyset$, the empty function.

[2] In this sense, the use of ZFC* in set theory is like the use of ε in basic analysis.
[3] Note that this differs from the use of the word incompatible for partial orders in the theory of orders.
[4] This is the Cohen forcing!

The desired object is then obtained by 'gluing together' various approximations. But since we may have approximations that do not agree with each other, we cannot simply take the union. For example $\bigcup \mathbb{P}$ above is not going to be a function, because p, the partial function which assigns 0 to 0 and q, the partial function which assigns 1 to 0 are both in $\bigcup \mathbb{P}$ but $p \cup q$ is not a function (i.e. p and q are incompatible). Therefore we do not take the union of the whole $\mathbb{P}$ but of some filter G. In G every two conditions are compatible, so at least we shall get a function as $\bigcup G$.

However, if we choose our G trivially, for example $G = \emptyset$, we may end up with a trivial function, $\bigcup G = \emptyset$ for example. Therefore we require our G to satisfy various requirements, which we express using the dense sets. For example, in the Cohen forcing, every set $D_n = \{p : n \in \mathrm{dom}(p)\}$ is dense for $n \in \omega$ (we invite the reader to check this), so if our generic G intersects all of these sets then $\bigcup G$ will be a total function from ω to 2. [5]

Is there however such a generic set? Yes, if there are only countably many dense sets to meet, as we see next.

Before that though, a point on notation. We have presented forcing using the original notation of [12]. Much work on forcing has been done using a later notation and view developed by Dana Scott and Robert M. Solovay (see a discussion in [VII, [63]]) which is very natural if one thinks of forcing in terms of complete Boolean algebras, a view that we shall not have the space to develop here. It turns out that the two approaches are equivalent and that the practical difference to the reader is minimal: in our notation $p \leq q$ means that q gives more information than p. In the other notation, including the one in [63] this is written as $q \leq p$. As this book will discuss at length some major references which use Cohen's notation (such as the book [105]) and since the author is a firm believer in that notation, which best expresses her intuition about forcing, she has made the choice here and in her other work to be using the notation in question.

Lemma 8.2.2 *Suppose that $\mathbb{P}$ is a forcing notion and D_n $(n < \omega)$ are dense sets in $\mathbb{P}$. Then there is a $\mathbb{P}$-filter G which intersects all D_n.*

Proof By induction on $n < \omega$ we choose p_n such that $p_{n+1} \in D_n$ and $p_n \leq p_{n+1}$. We start with p_0 arbitrary and then continue using the density of each D_n.

At the end we let $G = \{q \in P : \exists n\,(q \leq p_n)\}$. It is easy to check that G is as required. $\bigstar_{8.2.2}$

This is why we needed in the logic of forcing to work with countable models of ZFC*. Namely, use the convention that from now on V is the fixed model of

[5] This is quite similar to classical priority arguments in recursion (computability) theory.

ZFC provided by the assumption Con(ZFC) and Gödel's Completeness Theorem, and argue within V. (Since we are then within V we may as well assume that V is the universe $\mathbf{V}$ of all sets, as we defined it before.[6])

Corollary 8.2.3 *Suppose that M is a countable model of* ZFC* *and $\mathbb{P}$ is a forcing notion in M. Then (in V) there is a filter G which is $\mathbb{P}$-generic over M.*

Proof In V we can enumerate all sets D that are dense subsets of $\mathbb{P}$ as D_n ($n < \omega$), as M is countable. Then we can apply Lemma 8.2.2 to get a filter G which intersects all D_n, hence it is $\mathbb{P}$-generic over M. $\bigstar_{8.2.3}$

This argument does not work if we do not assume the countability of M. Here is an example:

Claim 8.2.4 Let $\mathbb{P}$ be the Cohen forcing described by the formula (8.1) above, and let g be any function in ${}^{\omega}2$. Then the set $D = \{p \in \mathbb{P} : \exists n\,(p(n) \neq g(n))\}$ is dense in $\mathbb{P}$.

Proof Let $p \in \mathbb{P}$ and let $n < \omega$ be such that $n \notin \text{dom}(p)$, which exists as p is finite. Let q be $p \cup \{(n, 1 - g(n))\}$, hence $q \geq p$ and $q \in D$. $\bigstar_{8.2.4}$

Corollary 8.2.5 *Suppose that M is a model of* ZFC* *which contains all elements of 2^{ω} (that is $(2^{\omega})^V$).*

Then there is no Cohen-generic filter over M in V.

Proof Suppose that G were such a filter. By the properties of a filter in Cohen forcing as discussed above, $f = \bigcup G$ is a function in ${}^{\omega}2$. However, as M contains all elements of 2^{ω} and models ZFC*, in M we can describe f as the characteristic function of the set $A = \{n : f(n) = 1\}$, so $f \in M$. But then D_f (as in Claim 8.2.4) is a dense subset of V and is in M. As G is Cohen-generic over M we have $G \cap D_f \neq \emptyset$, so let $p \in G \cap D_f$. So there is n such that $p(n) \neq f(n)$. This is a contradiction, since $p \in G \implies p \subseteq f$. $\bigstar_{8.2.5}$

Coming back to Corollary 8.2.3, our motivation is that the model $M[G]$, the generic extension, is going to be a model not equal to M, but at the same time as close to M as possible. In fact, we shall achieve that $M[G]$ is the smallest model containing M and G and having the same ordinals as M. Therefore we need to make sure that G is not in M if M and $M[G]$ are to be different. For this reason, we work with separative partial orders, defined below:

[6] We make a remark addressed at a philosophically minded reader: there are philosophers of mathematics who argue that mathematics does not have a universe but rather a multiverse obtained by various generic extensions. The above shows that to argue about the correctness of this view, it first has to be stated very carefully (as is done by some but not all proponents of this view), since to take any generic extensions to start with, we already need to start with a universe within which we do forcing.

Definition 8.2.6 A partial order P is said to be *separative* if for every $p \in P$ there are $q, r \geq p$ such that $q \perp r$.

For example, Cohen forcing is separative as for a given p we can extend it to $q = p \cup \{(\mathrm{dom}(p), 0)\}$ and $r = p \cup \{(\mathrm{dom}(p), 1)\}$, which are incompatible. The importance of separative partial orders is the following:

Claim 8.2.7 Suppose that $\mathbb{P}$ is a separative forcing notion and G is $\mathbb{P}$-generic over M, for some M a transitive model of ZFC*. Then $G \notin M$.

Proof Suppose otherwise. Let $D = \mathbb{P} \setminus G$. We claim that D is a dense subset of $\mathbb{P}$. So, given $p \in \mathbb{P}$, as $\mathbb{P}$ is separative, we can find $q \perp r$ both extending p. Not both q, r can be in G, as G is a filter, so at least one has to be in D, proving that D is dense.

Since $G \in M$ and M is a model of ZFC*, D is in M – as we can evaluate it in M using the operations $\cup, \setminus$ which are expressible in M and absolute by transitivity, and the parameter G which is in M. But then as G is $\mathbb{P}$-generic over M we conclude $G \cap D \neq \emptyset$, a contradiction. $\bigstar_{8.2.7}$

8.3 $M[G]$

Now we need to describe what $M[G]$ is as a set. Throughout, let M stand for a countable transitive model of ZFC* and $\mathbb{P}$ for a forcing notion which is an element of M.

We define a $\mathbb{P}$-name:

Definition 8.3.1 By induction on $\in$, we define objects that are $\mathbb{P}$-*names*: $\underset{\sim}{\sigma}$ is a $\mathbb{P}$-name iff $(\forall x \in \underset{\sim}{\sigma})\ x = (p, \underset{\sim}{\tau})$ for some $p \in \mathbb{P}$ and $\mathbb{P}$-name $\underset{\sim}{\tau}$.

This definition looks strange at first, but in fact its validity uses the fact that $\in$ is well founded and the fact that the empty set is vacuously a $\mathbb{P}$-name. Note that the above definition gives a proper class of $\mathbb{P}$-names. We will be interested in those $\mathbb{P}$-names that are in M, so note that the definition of what it means to be a $\mathbb{P}$-name is absolute for M, and hence the $\mathbb{P}$-names that are in M form exactly the M-class of $\mathbb{P}$-names.

If we are given a filter G on $\mathbb{P}$ we can associate its interpretation $\underset{\sim}{\tau}_G$ (sometimes also denoted as $[\underset{\sim}{\tau}]_G$) to every $\mathbb{P}$-name $\underset{\sim}{\tau}$, by using another inductive definition.

Definition 8.3.2 For a $\mathbb{P}$-name $\underset{\sim}{\tau}$ and a filter G in $\mathbb{P}$, we define *the evaluation* $\underset{\sim}{\tau}_G$ by induction:

$$\underset{\sim}{\tau}_G = \{\underset{\sim}{\sigma}_G : (\exists p \in G)(p, \underset{\sim}{\sigma}) \in \underset{\sim}{\tau}\}.$$

We can check for example that for every G we have $\emptyset_G = \emptyset$. Now we let

$$M[G] = \{\underset{\sim}{\tau}_G : \underset{\sim}{\tau} \text{ is a } \mathbb{P}\text{-name} \in M\}.$$

The following shows why $M[G]$ is a 'minimal' extension of M:

Lemma 8.3.3 *Suppose that N is a transitive model of* ZFC*, *$M \subseteq N$ and $G \in N$. Then $M[G] \subseteq N$.*

Proof Notice that the evaluations $\underset{\sim}{\tau}_G$ are absolute between $M[G]$ and N, and hence $M[G] \subseteq N$. $\bigstar_{8.3.3}$

We now need to show that $M \subseteq M[G]$. This is the case because we can to each $a \in M$ associate a special name $\check{a}$ such that $\check{a}_G = a$. Namely we define $\check{\emptyset} = \emptyset$ and

$$\check{a} = \{(p, \check{b}) : p \in \mathbb{P} \text{ and } b \in a\}.$$

We can easily see that $\check{a}_G = a$ for every $a \in M$, and moreover this does not depend on G. There is also a canonical name for the generic filter G: define

$$\underset{\sim}{\Gamma} = \{(p, \check{p}) : p \in \mathbb{P}\}.$$

Then we have $\underset{\sim}{\Gamma}_G = G$ (and this does not depend on G).

We can also easily check that $M[G]$ is transitive. We could also easily check some of the axioms of ZFC that we want to hold in $M[G]$, such as Pairing, Extensionality and Foundation. Rather than doing that, we shall go on to introduce the forcing relation, which we need in order to better understand $M[G]$ and show all the axioms of ZFC.

8.4 The Forcing Relation

The title of this section is misleading because the whole point of the Main Theorem of Forcing is to show that two independently defined forcing relations, $\Vdash$ and $\Vdash^*$ are the same. The former one is known as the forcing relation and the latter as the 'weak forcing', whose primary use is to, in conjunction with the Main Theorem of Forcing, show that $\Vdash$ is expressible as a relation in M. Once we prove that theorem, we can forget about the weak forcing, so the title is reflective of the common usage of $\Vdash$ as the only forcing relation.

Definition 8.4.1 Suppose that M is a model of ZFC*, $\mathbb{P} \in M$ a forcing notion and $\varphi(x_0, \ldots, x_{n-1})$ a formula of the language of set theory. Then for $p \in \mathbb{P}$ and names $\underset{\sim}{\tau}_0, \ldots, \underset{\sim}{\tau}_{n-1}$ we say

$$p \Vdash_{\mathbb{P}} {}^{7} \varphi(\underset{\sim}{\tau}_0, \ldots, \underset{\sim}{\tau}_{n-1})$$

if for every $\mathbb{P}$-generic filter G over M such that $p \in G$, we have that $M[G] \models \varphi([\underset{\sim}{\tau}_0]_G, \ldots, [\underset{\sim}{\tau}_{n-1}]_G)$.

This definition clearly depends on (all possible!) G and therefore is not expressible in M. This is an example of a 'semantic' definition, a definition that depends on interpretation of $\models$. We are now going to give a definition of $\Vdash^*$, which is an example of a 'syntactic' definition, one which only depends on the syntactic properties of formulas. It will be convenient to have the following notion:

Definition 8.4.2 Suppose that $\mathbb{P}$ is a forcing notion and $p \in \mathbb{P}$. A set $D \subseteq \mathbb{P}$ is *dense above* p if for all $q \geq p$ there is $r \geq q$ with $r \in D$.

Definition 8.4.3 Let $\mathbb{P}$ be a forcing notion, let $\varphi(x_0, \ldots, x_{n-1})$ be a formula of the language of set theory, let $\underset{\sim}{\tau}_0, \ldots, \underset{\sim}{\tau}_{n-1}$ be $\mathbb{P}$-names and $p \in \mathbb{P}$ a condition. By induction on the complexity of φ we define what it means for p to weak-force $\varphi(\underset{\sim}{\tau}_0, \ldots, \underset{\sim}{\tau}_{n-1})$, i.e. $p \Vdash^* \varphi(\underset{\sim}{\tau}_0, \ldots, \underset{\sim}{\tau}_{n-1})$.

1. $p \Vdash^* (\underset{\sim}{\tau}_0 = \underset{\sim}{\tau}_1)$ if for all $(s, \underset{\sim}{\sigma}) \in \underset{\sim}{\tau}_0$, $\{q : q \geq s \implies (\exists (s', \underset{\sim}{\sigma}') \in \underset{\sim}{\tau}_1)(q \geq s' \wedge q \Vdash^* (\underset{\sim}{\sigma} = \underset{\sim}{\sigma}'))\}$ is dense above p, and similarly with $\underset{\sim}{\tau}_0$ and $\underset{\sim}{\tau}_1$ interchanged.
2. $p \Vdash^* (\underset{\sim}{\tau}_0 \in \underset{\sim}{\tau}_1)$ if $\{q : (\exists (s', \underset{\sim}{\sigma}') \in \underset{\sim}{\tau}_1)(q \geq s' \wedge q \Vdash^* (\underset{\sim}{\sigma}' = \underset{\sim}{\tau}_0))\}$ is dense above p.
3. $p \Vdash^* (\varphi \wedge \psi)$ iff $p \Vdash^* \varphi$ and $p \Vdash^* \psi$.
4. $p \Vdash^* \neg\varphi$ iff there is no $q \geq p$ with $q \Vdash^* \varphi$.
5. $p \Vdash^* (\exists x)\varphi(x)$ iff $\{q : (\exists \underset{\sim}{\tau})\, q \Vdash^* \varphi(\underset{\sim}{\tau})\}$ is dense above p.[8]

The definition of $\Vdash^*$ lets us express the satisfaction relation in $M[G]$. Keeping the notation from the above:

Lemma 8.4.4 *For any formula* $\varphi(x_0, \ldots, x_{n-1})$ *and names* $\underset{\sim}{\tau}_0, \ldots, \underset{\sim}{\tau}_{n-1}$,

- *If* $p \in G$ *and* $p \Vdash^* \varphi(\underset{\sim}{\tau}_0, \ldots, \underset{\sim}{\tau}_{n-1})$ *then* $M[G] \models \varphi({\underset{\sim}{\tau}_0}^G, \ldots, {\underset{\sim}{\tau}_{n-1}}^G)$,
- *If* $M[G] \models \varphi(\underset{\sim}{\tau}_0^G, \ldots, \underset{\sim}{\tau}_{n-1}^G)$ *then for some* $p \in G$ *we have* $p \Vdash^* \varphi(\underset{\sim}{\tau}_0 \ldots, \underset{\sim}{\tau}_{n-1})$.

We have used the notation $\underset{\sim}{\tau}_i^G$ in place of $\underset{\sim}{\tau}_{i,G}$ to improve readability, and we shall use such notation in some future instances as well.

[7] We may omit subscript $\mathbb{P}$ if it is clear from the context. We sometimes write $\Vdash_{\mathbb{P}}$ in place of $\emptyset_{\mathbb{P}} \Vdash_{\mathbb{P}}$.

[8] We are using the fact that from the above clauses we can uniquely define the action of $\Vdash^*$ on the remaining binary connectives and the universal quantifier.

Proof (outline) The proof is of both parts simultaneously, using induction on the complexity of φ. For the atomic case (using also an induction on the $\in$-complexity of the names), let us as an example suppose that some $p \in G$ satisfies $p \Vdash^* (\underset{\sim}{\tau}_0 = \underset{\sim}{\tau}_1)$ and show that $M[G] \models ([\underset{\sim}{\tau}_0]_G \subseteq [\underset{\sim}{\tau}_1]_G)$, and hence $M[G] \models ([\underset{\sim}{\tau}_0]_G = [\underset{\sim}{\tau}_1]_G)$ by symmetry. In $M[G]$ let $a = \underset{\sim}{\sigma}_G \in [\underset{\sim}{\tau}_0]_G$, hence for some $q \in G$ we have $(q, \underset{\sim}{\sigma}) \in \underset{\sim}{\tau}_0$. By the definition of $\Vdash^*$, the set

$$D = \{r \geq p : r \geq q \implies (\exists (q', \underset{\sim}{\sigma}') \in \underset{\sim}{\tau}_0)(r \geq q' \wedge q' \Vdash^* (\underset{\sim}{\sigma} = \underset{\sim}{\sigma}'))\}$$

is dense above p. Let $s \in G$ be a common extension of p and q, and hence D is dense above s. We can therefore find $r \geq s$ with $r \in G \cap D$ and then we also have $r \geq q$. We have $q' \Vdash^* (\underset{\sim}{\sigma} = \underset{\sim}{\sigma}')$. By the induction hypothesis we have $\underset{\sim}{\sigma}_G = \underset{\sim}{\sigma}'_G$ and therefore by the definition of D, we have $a = \underset{\sim}{\sigma}_G = \underset{\sim}{\sigma}'_G \in [\underset{\sim}{\tau}_1]_G$.

The non-atomic cases of φ are actually easier. For a full proof see [63, VII, 3.5.2]. ★8.4.4

Now we are ready to state the the Main Theorem of Forcing. Like some other great theorems in mathematics (such as the Main Gap Theorem of [102]), the Main Theorem of Forcing states that a semantic and a, seemingly unrelated, syntactic notion agree. In this case these are the forcing and the weak forcing relations.

Theorem 8.4.5 (Main Theorem of Forcing) *Let M be a countable transitive model of ZFC*, $\mathbb{P}$ a forcing notion in M, $\varphi(x_0, \ldots, x_{n-1})$ a formula in the language of set theory, and $p \in \mathbb{P}$ a condition. Then for any $\mathbb{P}$-names $\underset{\sim}{\tau}_0, \ldots, \underset{\sim}{\tau}_{n-1}$ in M we have*

$$p \Vdash \varphi(\underset{\sim}{\tau}_0, \ldots, \underset{\sim}{\tau}_{n-1}) \iff (p \Vdash^* \varphi(\underset{\sim}{\tau}_0, \ldots, \underset{\sim}{\tau}_{n-1}))^M.$$

In particular the forcing relation $\Vdash_{\mathbb{P}}$ is expressible in M.

For the proof of the Main Theorem of Forcing see [63, VII, §3]. The main use of this theorem is in applications of its last sentence. We shall encounter such a use as soon as we reach the next chapter.

Corollary 8.4.6 *1. For any formula $\varphi(x_0, \ldots, x_{n-1})$ and names $\underset{\sim}{\tau}_0, \ldots, \underset{\sim}{\tau}_{n-1}$ we have $M[G] \models \varphi(\underset{\sim}{\tau}_0{}^G, \ldots, \underset{\sim}{\tau}_{n-1}G)$ iff for some $p \in G$, $p \Vdash \varphi(\underset{\sim}{\tau}_0, \ldots, \underset{\sim}{\tau}_{n-1})$.*
2. $\neg p \Vdash \varphi(\underset{\sim}{\tau}_0, \ldots, \underset{\sim}{\tau}_{n-1})$ iff $\{q : q \Vdash \neg\varphi(\underset{\sim}{\tau}_0, \ldots, \underset{\sim}{\tau}_{n-1})\}$ is dense above p.

Another important observation which becomes crucial in arguments in Part Two of the book is the following lemma.

Lemma 8.4.7 (Existential Completeness Lemma) *Suppose that $\mathbb{P}$ is a forcing notion, $p \in \mathbb{P}$ and φ is a formula. Then*

$$p \Vdash (\exists x)\varphi(x) \text{ iff there is a } \mathbb{P}\text{-name } \underset{\sim}{\tau} \text{ such that } p \Vdash \varphi(\underset{\sim}{\tau}).$$

Proof We know by 8.4.5 that $p \Vdash (\exists x)\varphi(x)$ iff $(p \Vdash^* (\exists x)\varphi(x))^M$, which by definition of weak forcing means that in M, the set $\mathcal{D} = \{q : (\exists \underset{\sim}{\tau}) : q \Vdash \varphi(\underset{\sim}{\tau})\}$ is dense above p. Note that $\mathcal{D} \in M$. Let us enumerate a maximal antichain in $\mathcal{D}$ as $\{q_i : i < \alpha^*\}$ for some α^* and for each i choose a name $\underset{\sim}{\tau}_i$ such that $q_i \Vdash \varphi(\underset{\sim}{\tau}_i)$. Now we define $\underset{\sim}{\tau} = \{(q_i, \underset{\sim}{\tau}_i) : i < \alpha^*\}$. If G is $\mathbb{P}$-generic with $p \in G$, then there is exactly one of q_i which is in G and hence $\underset{\sim}{\tau}_G = \{\underset{\sim}{\tau}_{iG}\}$ and (modulo renaming to get rid of the $\{\}$), $M[G] \models \varphi(\underset{\sim}{\tau}_G)$. So $p \Vdash \varphi(\underset{\sim}{\tau})$. The other direction is easy. $\bigstar_{8.4.7}$

8.5 ZFC in $M[G]$

Theorem 8.5.1 *With the notation of previous chapters, $M[G]$ satisfies* ZFC*, *i.e. it satisfies the same amount of* ZFC *as M does.*

Notation 8.5.2 For a name $\underset{\sim}{\sigma}$ let $\mathrm{dom}(\sigma)$ be the set of all $\underset{\sim}{\tau}$ such that $(p, \underset{\sim}{\tau}) \in \sigma$ for some p.

Proof (outline) We can easily check the axioms of Extensionality, Pairing, Union and Foundation. Let us check the Comprehension Scheme.

Let $\varphi(x, y, a_0, \ldots, a_{n-1})$ be a formula and let $\underset{\sim}{\sigma}$, $\underset{\sim}{\tau}$, $\underset{\sim}{\rho}_0, \ldots, \underset{\sim}{\rho}_{n-1}$ be $\mathbb{P}$-names in M. We need to verify that $A = \{a \in \underset{\sim}{\sigma}_G : \varphi(a, \underset{\sim}{\tau}_G, [\underset{\sim}{\rho}_0]_G, \ldots, [\underset{\sim}{\rho}_{n-1}]_G)\}$ is a set in $M[G]$. Therefore we need to find a name in M whose interpretation is this set. Let

$$\underset{\sim}{\rho} = \{(p, \underset{\sim}{\pi}) : p \in \mathbb{P},\ \underset{\sim}{\pi} \in \mathrm{dom}(\sigma) \wedge p \Vdash (\underset{\sim}{\pi} \in \sigma \wedge \varphi(\underset{\sim}{\pi}, \underset{\sim}{\tau}, \underset{\sim}{\rho}_0, \ldots, \underset{\sim}{\rho}_{n-1}))\}.$$

By the Main Theorem of Forcing, $\underset{\sim}{\rho}$ is in M (as $\Vdash$ is expressible in M). Let us see that $\underset{\sim}{\rho}_G = A$. Suppose that $a \in A$, hence by the definition of $\underset{\sim}{\sigma}_G$, $a = \underset{\sim}{\pi}_G$ for some $(p, \underset{\sim}{\pi}) \in \sigma$ such that $p \in G$. The pair $(p, \underset{\sim}{\pi})$ is easily seen to satisfy the first condition of being in $\underset{\sim}{\rho}$. Since $\underset{\sim}{\pi}_G \in A$ there must be a $q \in G$ which forces this, so $q \Vdash \varphi(\underset{\sim}{\pi}, \underset{\sim}{\tau}, \underset{\sim}{\rho}_0, \ldots, \underset{\sim}{\rho}_{n-1})$. Since G is a filter, we can find $r \in G$ with $r \geq p, q$ and hence $(r, \pi) \in \underset{\sim}{\rho}$ and therefore $a \in \underset{\sim}{\rho}_G$. It is clear that $\underset{\sim}{\rho}_G \subseteq A$, so we are done.

Now let us verify Replacement. Let us suppose that we are given a formula $\varphi(x, y, z, a_0, \ldots, a_{n-1})$, names $\underset{\sim}{\sigma}$, $\underset{\sim}{\rho}_0, \ldots, \underset{\sim}{\rho}_{n-1}$ in M and that

$$(\forall x \in \underset{\sim}{\sigma}_G)(\exists! y)\, \varphi(x, y, \underset{\sim}{\sigma}_G, [\underset{\sim}{\rho}_0]_G, \ldots, [\underset{\sim}{\rho}_{n-1}]_G)$$

holds in $M[G]$. For simplicity in notation let us suppress the parameters. By Replacement in M,[9] we can find a set $S \in M$ such that $(\forall \underset{\sim}{\pi} \in \mathrm{dom}(\underset{\sim}{\sigma}))(\forall p \in \mathbb{P})$

[9] applied by picking the $\underset{\sim}{\mu}$ of the lowest rank α and first in the well ordering of V_α within M

if there is $\underset{\sim}{\mu}$ in M such that $p \Vdash \varphi(\underset{\sim}{\pi}, \underset{\sim}{\mu})$ then there is such $\underset{\sim}{\mu}$ in S. Let $\underset{\sim}{\tau} = \{(\emptyset_{\mathbb{P}}, \underset{\sim}{\mu}) : \underset{\sim}{\mu} \in S\}$. Now we can check that $\underset{\sim}{\tau}_G$ satisfies that $(\forall x \in \underset{\sim}{\sigma}_G)(\exists y \in \underset{\sim}{\tau}_G)\varphi(x, y, \underset{\sim}{\sigma}_G, [\underset{\sim}{\rho}_0]_G, \ldots, [\underset{\sim}{\rho}_{n-1}]_G)$.

The Axiom of Infinity holds because ω is absolute. For the Power Set, notice that in general for $x \in M$ we do not have $\mathcal{P}(x)^M = \mathcal{P}(x)$, as M is countable, and the same will be true of $M[G]$. However, we only need to show that for every $\underset{\sim}{\sigma}_G \in M[G]$ we can find $\underset{\sim}{\mu}$ such that $\underset{\sim}{\mu}_G = \{y \subseteq \underset{\sim}{\sigma}_G : y \in M[G]\}$, and in fact since we have already proved that Comprehension holds, it suffices to produce $\underset{\sim}{\mu}$ such that $y \subset \underset{\sim}{\sigma}_G, y \in M[G] \implies y \in \underset{\sim}{\mu}_G$. Let S be the set of all $\mathbb{P}$-names $\underset{\sim}{\tau}$ in M such that $\mathrm{dom}(\underset{\sim}{\tau}) \subseteq \mathrm{dom}(\underset{\sim}{\sigma})$ and let $\underset{\sim}{\mu} = \{\emptyset^{\mathbb{P}}\} \times S$.

Finally we need to prove the Axiom of Choice. For this we shall show that for any $x \in M[G]$ there is in $M[G]$ a function from some ordinal α onto x. This suffices, because then we can define a well order on x by saying for $y, z \in x$ that $y < z$ iff $\min(f^{-1}(y)) < \min(f^{-1}(z))$. Let $x = \underset{\sim}{\tau}_G$. In M, by the Axiom of Choice, we can enumerate $\mathrm{dom}(\underset{\sim}{\tau}) = \{\underset{\sim}{\pi}_\gamma : \gamma < \alpha\}$ for some ordinal α. Then in $M[G]$ we can consider the function which to $\gamma < \alpha$ associates $(\underset{\sim}{\pi}_\gamma)_G$. $\bigstar_{8.5.1}$

8.6 Ordinals in $M[G]$

Claim 8.6.1 Suppose that M is a countable transitive model of ZFC*. Then there is a countable limit ordinal δ such that $\mathrm{Ord} \cap M = \delta$.

Proof By transitivity, if α is an ordinal and in M then $\alpha \subseteq M$. Since M is countable, there must be the first ordinal δ which is not in M, and by the above observation δ is countable and satisfies the assumptions of the claim. $\bigstar_{8.6.1}$

We denote the ordinal satisfying the conclusion of Claim 8.6.1 by $o(M)$.

Claim 8.6.2 $o(M[G]) = o(M)$.

Proof By induction on the rank, we can show that for any name $\underset{\sim}{\tau}$, the $M[G]$-rank of $\underset{\sim}{\tau}_G$ is $\leq$ the M-rank of $\underset{\sim}{\tau}$. In particular the rank of any ordinal α in $M[G]$ is α, and is $\leq$ the rank, say β, of $\underset{\sim}{\tau}$ for any $\underset{\sim}{\tau}$ in M with $\underset{\sim}{\tau}_G = \alpha$. Then $\beta \in M$ and hence by transitivity also $\alpha \in M$. $\bigstar_{8.6.2}$

8.7 What is New in $M[G]$?

Theorem 8.7.1 *If $G \notin M$ then $M[G]$ does not satisfy the Axiom of Constructibility.*

Proof We have that by $o(M) = o(M[G])$ also $\mathbf{L}^{M[G]} = \mathbf{L}^M \neq M[G]$. $\bigstar_{8.7.1}$

9
Violating CH

Forcing was invented to prove the independence of CH. We have already indicated that this is going to be done through adding 'new' subsets of ω, to the extent that we shall have more than ω_1 of them. What do we mean by ω_1 here? We know that whatever $M[G]$ is going to understand by ω_1 is simply some countable ordinal (as any ordinal that is in $M[G]$ is countable) so what we are aiming for here is that $M[G] \models (2^\omega > \omega_1)$, so $(\omega_1)^{M[G]}$, i.e. the ω_1 that we mean here is the first ordinal in $M[G]$ which is not countable from the point of view of $M[G]$. This means that there is no function in $M[G]$ which surjectively maps ω onto ω_1. This discussion contains the proof of the following:

Claim 9.0.1 $\omega_1^{M[G]} \geq \omega_1^M$.

Proof If $\alpha \in M$ is such that $M \models$ (α is countable), then this means that in M there is a surjection from ω onto α. The same surjection will witness that α is countable in $M[G]$. $\bigstar_{9.0.1}$

Similar statements may be made about other cardinals. We say:

Definition 9.0.2 A forcing notion $\mathbb{P} \in M$ *preserves cardinalities* if for every α which is a cardinal in M, α remains a cardinal in $M[G]$, for any $\mathbb{P}$-generic filter G over M.

In particular, if $\mathbb{P}$ preserves cardinalities then $\omega_1^{M[G]} = \omega_1^M$, $\omega_2^{M[G]} = \omega_2^M$ etc. To produce a model in which CH fails, we shall start with a model M of CH and force with a forcing notion that preserves cardinals and makes $2^\omega \geq \omega_2^M = \omega_2^{M[G]}$. The forcing notion is

$$\mathbb{P} = \{p\colon \omega \times \omega_2 \to 2 : \mathrm{dom}(p) \text{ is finite}\},$$

ordered by inclusion. We can think of this forcing as being made up of ω_2 independent copies of Cohen forcing. It is usually referred to as 'adding ω_2

Cohen reals'. The reason is that if we denote by G the generic of this forcing, then G is a function from $\omega \times \omega_2 \to 2$ and simple density arguments show that G is a total function. In $M[G]$ we can define for $\alpha < \omega_2$ a function $f_\alpha \colon \omega \to 2$ by letting $f_\alpha(n) = G(n, \alpha)$.

Lemma 9.0.3 *Whenever $\alpha \neq \beta < \omega_2$ we have that $f_\alpha \neq f_\beta$.*

Proof We note that the set

$$D = \{p : (\exists n)((n, \alpha), (n, \beta) \in \mathrm{dom}(p),\ p((n, \alpha)) \neq p((n, \beta)))\}$$

is in M and dense in $\mathbb{P}$. $\bigstar_{9.0.3}$

This forcing is normally denoted by $\mathrm{Add}(\omega, \omega_2)$ and we can similarly define forcing $\mathrm{Add}(\kappa, \lambda)$ which adds λ subsets to κ by forcing with functions from $\lambda \times \kappa$ to 2 whose domains have size $< \kappa$.

The only point that remains to prove to finish the proof that in Cohen's model the value of the continuum is at least $\aleph_2$, is that the forcing preserves cardinalities. This will be done through an analysis of the antichains in the forcing.

In fact, it turns out that rather than using all the names, to obtain $M[G]$ it suffices to use rather simple names controlled by the antichains of the forcing, which we study in §9.2. This is what will allow us to calculate the exact value of the continuum in the Cohen model.

9.1 Chain Conditions and Closure

Here are two basic tools in the study of preservation of cardinalities.

Definition 9.1.1 Suppose that κ is an infinite cardinal and $\mathbb{P}$ is a forcing notion.

(1) We say that $\mathbb{P}$ has *κ-chain condition* if every antichain in $\mathbb{P}$ has size $< \kappa$. The property of being $\aleph_1$-cc is called *ccc*.

(2) We say that $\mathbb{P}$ is $(< \kappa)$-closed if every increasing sequence of length $< \kappa$ in $\mathbb{P}$ has an upper bound.

The point of these notions is that κ^+-cc forcing preserves the cardinalities and the cofinalities $> \kappa$ and $(< \kappa)$-closed forcing preserves cardinalities and cofinalities $\leq \kappa$, as we now show. We shall use the terminology *the ground model* to refer to the model over which we force.

Lemma 9.1.2 *(1) Suppose that $\mathbb{P}$ is a κ^+-cc forcing. Then for every two ordinals α, β and a function $f : \alpha \to \beta$ in the extension by $\mathbb{P}$, there is g in the ground model such that $g : \alpha \to [\beta]^{\leq\kappa}$ with $\Vdash_{\mathbb{P}}$ '$(\forall i < \alpha) f(i) \in g(i)$'.*
(2) Suppose that $\mathbb{P}$ is a $(< \kappa)$-closed forcing. Then for every two ordinals α, β and a function $f : \alpha \to \beta$ in an extension by $\mathbb{P}$, if $\alpha < \kappa$, then f belongs to the ground model.
(3) Consequently, κ^+-cc forcing preserves the cardinalities and the cofinalities $> \kappa$ and $(< \kappa)$-closed forcing preserves cardinalities and cofinalities $\leq \kappa$.

Proof (1) Let $\underset{\sim}{f}$ be a $\mathbb{P}$-name for f. For every $i < \alpha$, consider

$$\mathcal{D}_i = \{p : p \text{ forces a value to } \underset{\sim}{f}(i)\}.$$

Then $\mathcal{D}_i$ is a dense set in $\mathbb{P}$. Let $\mathcal{A}_i$ be a maximal antichain within $\mathcal{D}_i$ (which exists by Zorn Lemma). Notice that by the density of $\mathcal{D}_i$, the antichain $\mathcal{A}_i$ is maximal in $\mathbb{P}$, and hence every generic filter will contain an element of $\mathcal{A}_i$. Let $g(i) = \{j : (\exists p \in \mathcal{A}_i)\, p \Vdash \text{'}\underset{\sim}{f}(i) = j\text{'}\}$, hence $\Vdash_{\mathbb{P}}$ '$\underset{\sim}{f}(i) \in g(i)$'. We have that $|g(i)| \leq |\mathcal{A}_i| \leq \kappa$.
(2) Let $\underset{\sim}{f}$ be a $\mathbb{P}$-name for a function from $\alpha < \kappa$ to β. By extending the function in some trivial way, we can assume that α is a limit ordinal. We shall show that the set of all conditions p such that p forces that $\underset{\sim}{f} = g$ for some ground model function g is dense, which suffices for the conclusion.

Given $p_0 \in \mathbb{P}$. By induction on $i < \alpha$ we choose an increasing sequence of conditions $\langle p_i : i < \alpha \rangle$ so that p_{i+1} decides a value of $\underset{\sim}{f}(i)$, which we shall call $g(i)$. The induction is straightforward, as at limit stages $\delta < \alpha$ we can use the $(< \kappa)$-closure to find p_δ which is above all p_i for $i < \delta$. At the end of this induction, we can use the closure one more time to obtain a common extension p^* to all p_i for $i < \alpha$. Then $p^* \Vdash$ '$\underset{\sim}{f} = g$'.
(3) For the preservation by κ^+-cc forcing: suppose that $\mathrm{cf}(\alpha) > \kappa$ holds in the ground model but for some $\lambda \leq \kappa$ there is a function from $f \in M[G]$ which is cofinal from λ to α. Let g be as provided by (1) and for each $i < \lambda$ let $h(i) = \sup(g(i))$. By the assumption we have that each $h(i) < \alpha$, but putting the properties of $\underset{\sim}{f}$ and g together, we obtain that $h : \lambda \to \alpha$ is cofinal, a contradiction.

For the preservation by $(< \kappa)$-closed forcing: suppose that $\theta = \mathrm{cf}(\alpha) \leq \kappa$ holds in the ground model but for some $\lambda < \theta$ there is a function from $f \in M[G]$ which is cofinal from λ to α. By (2), that function is actually in M, a contradiction. $\bigstar_{9.1.2}$

Note that preserving all cofinalities (what this means is defined in the obvious way) immediately implies preserving all cardinalities. We now discuss the chain condition and the closure properties of the Cohen forcing. We shall often use the following well known combinatorial fact. It seems to have been discovered many times independently, but the earliest reference we could find is due to Nikolai A. Shanin in [98].

Definition 9.1.3 A family $\mathcal{A}$ of sets forms a Δ-*System* if there is a set r, called *the root* such that for all $A \neq B$ from $\mathcal{A}$ we have $A \cap B = r$.

Lemma 9.1.4 (Δ-System Lemma) *Suppose that $\lambda < \kappa$ are regular cardinals and that $(\forall \theta < \kappa)\ \theta^{<\lambda} < \kappa$. Then for every family of κ many sets each of cardinality $< \lambda$, there is a subfamily of size κ which forms a Δ-System.*

There are several proofs of the Δ-System Lemma, the one that seems most elegant while remaining combinatorial is in [II 1.5 [63]].

9.2 Canonical Names, again

Definition 9.2.1 [1] A name $\underset{\sim}{\tau}$ is a *canonical name* for a subset of $\underset{\sim}{\sigma}$ for some name $\underset{\sim}{\sigma}$ if it is of the form $\bigcup\{A_{\underset{\sim}{\pi}} \times \{\underset{\sim}{\pi}\} : \underset{\sim}{\pi} \in \mathrm{dom}(\underset{\sim}{\sigma})\}$.

The interest in canonical names is that on the one hand we can easily count them, yet on the other hand they are enough to provide us with all the necessary information. Namely, the following is true.

Lemma 9.2.2 *For any forcing notion $\mathbb{P} \in M$, object $\sigma \in M$ and a $\mathbb{P}$-name $\underset{\sim}{\mu}$ in M, there is a canonical name $\underset{\sim}{\tau}$ for a subset of σ such that*

$$\Vdash_{\mathbb{P}} `\underset{\sim}{\mu} \subseteq \sigma \implies \underset{\sim}{\mu} = \underset{\sim}{\tau}'.$$

Counting the canonical names is a technique that we can use to provide upper bounds for the cardinal arithmetic in the extension. For example, if we have a forcing notion $\mathbb{P}$ which has a κ-chain condition, then there are at most $|\mathbb{P}|^{<\kappa}$ antichains in $\mathbb{P}$, which gives a bound on the number of canonical names. For example, one can see that there are $\aleph_2$-canonical names in $\mathrm{Add}(\omega, \omega_2)$ for a subset of ω, from which one concludes that the value of $2^{\aleph_0}$ in the Cohen's extension is exactly $\aleph_2$.

[1] What we are going to call canonical names is called *nice names* in the classical reference [63], as canonical names refer to the names of the type $\check{a}$. Yet, the usage of the term canonical names seems to be more common now, hence our choice of terminology.

PART TWO

WHAT IS NEW IN SET THEORY

10
Introduction to Part Two

In this part we discuss results in set theory that have been obtained since the invention of forcing and up to now. So, not everything in this part is exactly new, but it goes beyond the popular perception of what set theory is, which seems to stop at Cohen's work. That is where this part of the book starts.

It is certainly impossible to include everything new in set theory in this part, but we presume that the reader is mostly interested to know what has happened in the theory of forcing. Hence we concentrate on the developments in that area. Again, we are going to be far from all-inclusive, but we shall give a consistent story. We concentrate on the developments in combinatorial set theory as the focus of the book, but we warn the reader that an entire forcing book could be written on the connections between determinacy, inner model theory and forcing–all of which are missing here.

The reader will notice that our notation changes to address the ground model most often as V (although we keep M for the proof of the consistency of Martin's Axiom in order for the reader to easily go back and forth between our text and the presentation in [63]). This is both historical and practical, since the letter M is used for some other notions in the theory of proper forcing and side conditions, for example, as we shall see in the relevant chapters. The reader who has made it so far has clearly entered a club of those who can tolerate changes of notation and imprecisions that one finds in research papers as opposed to classical textbooks, so we will not prolong the text of the rest of the book with such trivial remarks.

11
Classical Extensions

In this short chapter we discuss some forcing notions that although not that much newer than Cohen forcing, already form part of an advanced set theory investigation. Yet, these are so classical that one should certainly know them even before thinking of the advances that came later. In addition to the forcing notions studied in this chapter, another one is Easton forcing, given in §17.1, and yet another the notion of random reals which we let the reader discover from [51] or [63].

11.1 Lévy Collapse

The Lévy Collapse is the name of a forcing notion discovered by Azriel Lévy in [69]. Its purpose is to collapse a cardinal to become $\aleph_1$ in the extension and it is usually used in the context of collapsing a large cardinal, although we shall use $\mathrm{Col}(\aleph_1, \aleph_2)$ in the proof of Theorem 15.5.2.

Definition 11.1.1 Suppose that $\kappa > \aleph_1$ is regular and define $\mathrm{Col}(\aleph_1, \kappa)$, also denoted $\mathrm{Col}(\omega_1, \kappa)$, to be the set

$$\{p : p \text{ a finite partial function on } \kappa \times \omega \text{ and } (\forall (\alpha, n) \in \mathrm{dom}(p))\ p((\alpha, n)) \in \alpha\},$$

ordered by inclusion.

The result of forcing with $\mathrm{Col}(\aleph_1, \kappa)$ is to make all ordinals $< \kappa$ countable, since for every $\alpha < \kappa$ the function

$$f_\alpha(n) = p(\alpha, n) \text{ for some (any) } p \in G \text{ with } (\alpha, n) \in \mathrm{dom}(p)$$

is a generic surjection of ω onto α, as can be seen by producing the appropriate dense sets. The cardinal κ and the cardinals above it are preserved because of the following lemma.

Lemma 11.1.2 *For $\kappa \geq \aleph_1$ regular, the forcing* $\mathrm{Col}(\aleph_1, \kappa)$ *has κ-cc.*

Proof Suppose that $\langle p_\zeta : \zeta < \kappa \rangle$ is a sequence of conditions from $\mathrm{Col}(\aleph_1, \kappa)$. By Δ-System Lemma, passing to a subsequence of length κ if necessary, we can assume that $\{\mathrm{dom}(p_\zeta) : \zeta < \kappa\}$ form a Δ-System with root r^*. Let

$$\alpha^* = \max\{\alpha : (\exists n)(\alpha, n) \in r^*\},$$

so $\alpha^* < \kappa$. We now have that for all $\zeta < \kappa$, the function $p_\zeta \upharpoonright r^*$ is a finite function to α^*, so since $\alpha^* < \kappa$, we can without loss of generality assume that all $p_\zeta \upharpoonright r^*$ are the same function. Hence any two p_ζ, p_ξ are compatible in $\mathrm{Col}(\aleph_1, \kappa)$ and a common extension is $p_\zeta \cup p_\xi$. $\bigstar_{11.1.2}$

In analogy with the above, we can define $\mathrm{Col}(\lambda^+, \kappa)$, whose elements are partial functions from $\kappa \times \lambda^+$ of size $< \lambda$ and which satisfy that $p(\alpha, \zeta) \in \alpha$, ordered by inclusion. This forcing is $(< \lambda)$-closed and in the usual situation that κ is strongly inaccessible it has κ-cc (by an easy modification of the proof of Lemma 11.1.2), so it preserves cardinals up to and including λ and the ones above and including κ, forcing κ to become λ^+. Similarly we can define forcing of the type $\mathrm{Col}(< \lambda, \kappa)$.

11.2 Solovay Model

Another classic of the theory of forcing is a model due to Solovay [112] in which all sets of reals are Lebesgue measurable. This implies that the Axiom of Choice does not hold, since Giuseppe Vitali proved in [124] that the Axiom of Choice implies the existence of a subset of the real line which is not Lebesgue measurable. However, a limited amount of Choice does hold in the model, namely the *Axiom of Dependent Choice* (DC) which allows countably many consecutive choices. Solovay's model satisfies many other interesting regularity properties for sets of real numbers, such as the fact that every subset of the reals contains a perfect set or has the property of Baire. We shall not define these properties, assuming that the reader already knows them.

Solovay starts with a model V of ZFC in which κ is strongly inaccessible and uses $\mathrm{Col}(\omega_1, \kappa)$ to force κ to become ω_1 in the extension $V[G]$. Then $V[G]$ is still a model of ZFC and then, of course, not all subsets of $\mathbb{R}$ in it are measurable, by the result of [124]. However, Solovay passes to a class submodel HOD, of hereditarily ordinal definable sets. The proper definition of HOD takes a bit of work and can be found in [V, [63]].

In Solovay's model $V[G]$, every subset of $\mathbb{R}$ which is in $\mathrm{HOD}(\mathcal{P}(\omega))$ is

Lebesgue measurable and has the property of Baire, while HOD($\mathcal{P}(\omega)$) is a model of ZF and DC.

After Solovay's result it was asked if large cardinals were needed to obtain a model of ZF in which all sets of reals are Lebesgue measurable and have the property of Baire. It had been a long established experimental truth that the Lebesgue measure and the property of Baire enjoy a certain duality (see for example John Oxtoby's book [87]) and nobody had even imagined that the answer to the 'Solovay's inaccessible question' could be different between the Lebesgue measurability and the property of Baire. Therefore, the answer provided by Shelah in [101] came as quite a surprise: Shelah proved that for obtaining a model in which every set of reals is Lebesgue measurable, an inaccessible is needed, while there is no need for an inaccessible in obtaining a model in which every set of reals has the property of Baire.

12
Iterated Forcing and Martin's Axiom

In this chapter we see how the technique of forcing may be used to obtain many more independence results, involving important problems such as the Souslin problem. We see our first forcing axiom, Martin's Axiom.

12.0.1 Doing Forcing

We have described the technique of Cohen forcing and have noticed that an important point in his argument is that when passing from M to $M[G]$, the cardinals are preserved. This is true because the forcing used satisfies the countable chain condition. Cohen forcing for adding ω_2 many Cohen reals has an additional nice feature: it manages to do ω_2 many basic Cohen forcings at once, so it adds ω_2 many Cohen reals. It turns out that adding many objects at once, iterating the process of forcing, is not always such a simple process. For example, consider the idea of iterating forcing first from the semantic point of view:

Suppose we start with $M = M_0$, extend by G_0 which is $\mathbb{P}_0$-generic over M_0 and obtain $M_1 = M[G_0]$. Then we pick $\mathbb{P}_1 \in M_1$ and extend by G_1 which is $\mathbb{P}_1$-generic over M_1, obtaining $M_2 = M_1[G_1] = M[G_0][G_1]$, and so on. This is all reasonable until we reach the stage ω and then, what do we do? The most natural idea is to take $M_\omega = \bigcup_{n<\omega} M_n$. But examples exists to show that such an M_ω will not even contain the sequence $\langle G_n : n < \omega\rangle$ (see [63] Exercise VII. B.8), and hence is certainly not a generic extension by anything that we might understand as an iteration of the $\mathbb{P}_n$'s.

A brilliant idea, due to Solovay and Stanley Tennenbaum in [114] is to control this process all within M_0. So, to define in M_0 a forcing notion which encodes all the future extensions. In fact, the key idea is already present in the iteration of two steps of forcing. Rather than doing forcing with $\mathbb{P}_0$ in M_0 and then with $\mathbb{P}_1 \in M_1$, where $\mathbb{P}_1$ is chosen independently of $\mathbb{P}_0$ in some sort of a product $\mathbb{P}_0 \times \mathbb{P}_1$, the *iteration* of two forcing notions is defined in the following way.

Definition 12.0.1 Suppose that M is a countable transitive model of ZFC^*, $\mathbb{P}_0 \in M$ is a forcing notion and $\underset{\sim}{\mathbb{P}}_1$ is a $\mathbb{P}_0$-name for a forcing notion. Then the iteration $\mathbb{P}_0 * \underset{\sim}{\mathbb{P}}_1$ is a partial order defined on the set

$$\{(p, \underset{\sim}{q}) : p \in \mathbb{P}_0, p \Vdash_{\mathbb{P}_0} (\underset{\sim}{q} \in \underset{\sim}{\mathbb{P}}_1)\}$$

by the relation $(p, \underset{\sim}{q}) \le (p', \underset{\sim}{q}')$ iff $p \le_{\mathbb{P}_0} p'$ and $p' \Vdash_{\mathbb{P}_0} \underset{\sim}{q}' \ge \underset{\sim}{q}$.

For notational simplicity, it is usual to denote the elements of an iterated forcing $\mathbb{P}_0 * \underset{\sim}{\mathbb{P}}_1$ as pairs (p, q), although in reality the second coordinate is a name. Various properties may be proven about the iteration of two forcing notions, see Chapter VIII in [63], including the one that amounts to saying that forcing with $\mathbb{P}_0 * \underset{\sim}{\mathbb{P}}_1$ is equivalent to first forcing with $\mathbb{P}_0$ and then with the interpretation of $\underset{\sim}{\mathbb{P}}_1$ in the resulting $\mathbb{P}_0$-generic model. In particular, to any $\mathbb{P}_0 * \underset{\sim}{\mathbb{P}}_1$-generic H, we can associate a $\mathbb{P}_0$-generic G_0 over M and a $\mathbb{P}_1$-generic G_1 over $M[G_0]$ such that

$$M[H] = M[G_0][G_1], \tag{12.1}$$

and vice versa, from a $\mathbb{P}_0$-generic G_0 over M and a $\mathbb{P}_1$-generic G_1 over $M[G_0]$ we can define a $\mathbb{P}_0 * \underset{\sim}{\mathbb{P}}_1$-generic H satisfying (12.1) The property of iterated forcing that interests us most for the moment is the following:

Lemma 12.0.2 *Suppose that $\mathbb{P}_0 \in M$ is a ccc forcing notion and $\Vdash_{\mathbb{P}_0} (\underset{\sim}{\mathbb{P}}_1$ is a ccc forcing notion). Then $\mathbb{P}_0 * \underset{\sim}{\mathbb{P}}_1$ is ccc.*

Proof Suppose that $\langle (p_\alpha, q_\alpha) : \alpha < \omega_1 \rangle$ is an antichain in $\mathbb{P}_0 * \underset{\sim}{\mathbb{P}}_1$ and let G be $\mathbb{P}_0$-generic over M. Define $A \in M[G]$ as $A = \{\alpha : p_\alpha \in G\}$. Notice that for every $\alpha \ne \beta \in A$, we have that q_α, q_β are incompatible in $\underset{\sim}{\mathbb{P}}_1[G]$, as otherwise we would have that $(p_\alpha, q_\alpha), (p_\beta, q_\beta)$ are compatible, since G is a filter. On the other hand, we have that $\underset{\sim}{\mathbb{P}}_1[G]$ is ccc. Therefore, the set A must be countable in $M[G]$. Since G is arbitrary, we have that $\Vdash_{\mathbb{P}_0} (\exists \alpha)(\alpha < \omega_1$ and $\underset{\sim}{A} \subseteq \alpha)$.

By extending each p_α if necessary, we can assume that for each each $\alpha < \omega_1$ there is $\zeta_\alpha < \omega_1$ such that $p_\alpha \Vdash_{\mathbb{P}_0} (\sup(\underset{\sim}{A}) = \zeta_\alpha)$. Note that necessarily $\zeta_\alpha \ge \alpha$, since $p_\alpha \Vdash \alpha \in A$. Let $B \in [\omega_1]^{\aleph_0}$ be such that $\{p_\alpha : \alpha \in B\}$ is a maximal antichain among $\{p_\alpha : \alpha < \omega_1\}$, which exists since $\mathbb{P}_0$ is ccc. Let

$$\zeta^* = \sup\{\zeta_\alpha : \alpha \in B\},$$

so $\zeta^* < \omega_1$. Since each p_α is compatible with an element of B and each element of B forces that $\underset{\sim}{A} \subseteq \zeta^*$, we have that $\zeta_\beta \le \zeta^*$ for all $\beta < \omega_1$. Yet, taking $\beta \in (\zeta^*, \omega_1)$, we obtain a contradiction with $\zeta_\beta \ge \beta$. ★12.0.2

The above proof works similarly for proving that iteration of two factors preserves the property of κ-cc for any κ regular. An obvious generalisation of the above definitions gives an iteration with finitely many steps. However, if we wish to iterate more than finitely many times, we run into the problem of how to define the corresponding forcing notion and even more so, how to assure that cardinals are preserved by the iteration. Various techniques of iteration and corresponding iteration proofs form a vast area of research, on which a major reference is Shelah's book [105]. But, for ccc forcing a strong iteration theorem was discovered in [114].

Definition 12.0.3 A *finite support iteration* of forcing of length α^* is defined by induction on α^* as a sequence of the form

$$\langle P_\alpha, \underset{\sim}{Q}_\beta : \alpha \leq \alpha^*, \beta < \alpha^* \rangle,$$

satisfying:

- $P_0 = \{\emptyset\}$ is the trivial forcing notion,
- for every $\alpha < \alpha^*$, $\underset{\sim}{Q}_\alpha$ is a P_α-name of a forcing notion,
- P_α is a forcing notion with the order denoted as $\leq_\alpha$,
- $P_{\alpha+1} = P_\alpha * \underset{\sim}{Q}_\alpha$, for every $\alpha < \alpha^*$,
- for every limit $\delta \in (0, \alpha^*]$,

$$P_\delta = \{p_\delta = \langle p_i : i < \delta \rangle : (\forall i < \delta)\, p_\delta \restriction i \in P_i \text{ and } \\ \{i < \delta : \neg(p_\delta \restriction i \Vdash_{P_i} [p_i = \emptyset_{P_i}])\} \text{ is finite}\},$$

is a forcing notion ordered by $p_\delta \leq_\delta q_\delta$ iff for every $i < \delta$, $p_\delta \restriction i \leq_i q_\delta \restriction i$.

For a condition $p_\delta = \langle p_i : i < \delta \rangle$ in an iterated forcing, the set

$$\{i < \delta : \neg(p_\delta \restriction i \Vdash_{P_i} \text{`}p_i = \emptyset_{P_i}\text{'})\}$$

is called the *support* of p_δ and denoted by $\mathrm{supt}(p_\delta)$. By varying the requirement on the kind of support in an iterated forcing, we obtain various kinds of iterations. Among the ones that we shall see in this book are iterations with countable support (Definition 14.1.5), Easton support (Theorem 17.1.1) and $(<\kappa)$-support (Section 16.1).

In a forcing iteration $\mathbb{P} = \langle P_\alpha, \underset{\sim}{Q}_\beta : \alpha \leq \alpha^*, \beta < \alpha^* \rangle$, if $\varepsilon < \alpha \leq \alpha^*$, we use the notation $q \restriction \varepsilon$ for $q \in P_\alpha$ for the image of q under the natural projection of P_α to P_ε. In these circumstances, there is also a natural embedding of P_ε into P_α in which a condition $q \in P_\varepsilon$ is mapped to the unique condition $q' \in P_\alpha$ such that $q' \restriction \varepsilon = q$ and $q'(\beta) = \emptyset_{\underset{\sim}{Q}_\beta}$ for $\beta \in [\varepsilon, \alpha)$. In this case we write $q' = q \frown \emptyset_\varepsilon^\alpha$.

The main point about the iteration with finite support is the following theorem of [114], proved by methods similar to those of the proof of Lemma 12.0.2.

Theorem 12.0.4 *Suppose that $\langle P_\alpha, \underset{\sim}{Q}_\beta : \alpha \le \alpha^*, \beta < \alpha^* \rangle$ is a finite support iteration such that for all $\alpha < \alpha^*$, we have $\Vdash_{P_\alpha}$ '$\underset{\sim}{Q}_\alpha$ is ccc'. Then P_{α^*} is ccc.*

We shall not prove Theorem 12.0.4 now, since one can develop the theory of forcing axioms quite far without knowing this proof. Finally we shall give a more involved version of the proof in the proof of Theorem 16.1.3, from which the reader can easily go back and discover the proof of Theorem 12.0.4. Let us just say that a main tool used to sustain ccc is the Δ-System Lemma, which in the proof of Theorem 16.1.3 enters through its stronger version given in Claim 16.1.4.

Using Theorem 12.0.4 as a tool, Solovay and Tennenbaum obtained the consistency of the Souslin Hypothesis (there are no Souslin trees) with the negation of CH. The idea is to start with a model of CH and then to iterate with finite support ccc forcings which at each coordinate add an uncountable branch to a Souslin tree. A Souslin tree is an object of size $\aleph_1$ and one can show using combinatorial arguments that the bookkeeping can be arranged so that after ω_2 steps of iteration all candidates for a Souslin tree in the final extension have been taken care of and ω_2 subsets of ω have been added. A crucial point is the following observation, proved using an analysis of names:

Lemma 12.0.5 *If $\langle P_\alpha, \underset{\sim}{Q}_\beta : \alpha \le \omega_2, \beta < \omega_2 \rangle$ is an iteration of forcing with finite or countable supports and $S \in M$, then for every P_{ω_2}-name $\underset{\sim}{\tau}$ for a subset of S of size $\aleph_1$, there is $\alpha < \omega_2$ and a P_α-name $\underset{\sim}{\sigma}$ such that $\Vdash_{P_{\omega_2}} (\underset{\sim}{\tau} = \underset{\sim}{\sigma})$.*

In fact, a more general version of this lemma is true, see Lemma VIII 5.14 in [63], but the above instance is sufficient for what we need here and in the following sections.

12.1 Martin's Axiom

The famous lay mathematical reader, who is carefully reading this book on the plane or in between the administrative chores at some committee, might at this point find that the iteration of forcing looks useful but very technical and that after all, he is not going to spend his time learning the 'name' business. We are happy to report that maybe he does not have to.

As stated in [114], when Donald Martin saw the Solovay–Tennenbaum proof of the consistency of $SH + \neg$CH, he realised that the model obtained at the end of their iteration is *saturated* for generic extensions of the type they used. In fact, the individual forcings in the construction are formed of individual Souslin trees considered as partial orders. Each such forcing is ccc since the tree has no uncountable antichain. The forcing adds an uncountable branch to the tree. The saturation meant here is in fact Lemma 12.0.5 restated in these concrete circumstances: for every Souslin tree forcing $\mathbb{P}$ in the final model M_{ω_2} and every family $\mathcal{F}$ of $\aleph_1$ many dense sets in $\mathbb{P}$, M_{ω_2} contains an $\mathcal{F}$-generic filter. Martin realised that a more general theorem could be proved using exactly the same method. Namely, let us define Martin's Axiom:

Definition 12.1.1 $MA(\kappa)$ for a cardinal κ states that for every ccc forcing $\mathbb{P}$ and a family $\mathcal{F}$ of κ many dense sets in $\mathbb{P}$, there is an $\mathcal{F}$-generic filter.

It is easily seen that $MA(\omega)$ is simply true, $MA(2^{\aleph_0})$ is false, and varying the proof of Theorem 12.0.4 to consider all ccc forcing of size κ in an iteration of length κ^+ over a model of GCH, will give us $MA(\kappa)$ for ccc forcing of size κ. We obtain:

Theorem 12.1.2 *If ZFC is consistent then so is ZFC and the statement* $MA(\omega_1) + 2^{\aleph_0} = \aleph_2$.

Moreover, starting from a model of ZFC in which κ *is a regular uncountable cardinal satisfying* $2^{<\kappa} = \kappa$, *there is a ccc forcing which forces* $2^{\aleph_0} = \kappa$ *and* $MA(\lambda)$ *for all* $\lambda < \kappa$.

One important point about the consistency proof of $MA(\aleph_1) + \neg\mathrm{CH}$ is that it is obtained through an iteration of length ω_2 of ccc forcing of size $\aleph_1$ and the reason that the bookkeeping is possible is that $MA(\omega_1)$ holds iff it holds when restricted to partial orders of size $\aleph_1$, see Lemma VIII 6.2. in [63]. (Similar conclusion holds for other values of κ and $MA(\kappa)$.) We shall see in §15.5 that the analogue is not true when ccc forcing is replaced by the more general notion of proper forcing, and that it is this failure of the analogue that gives a large cardinal strength to the proper forcing axiom PFA, also to be discussed in §15.5.

To see how easy it is to use Martin's Axiom, let us prove the following theorem, due to Solovay–Tennenbaum and Martin, as explained above.

Theorem 12.1.3 *Under the assumption of* $MA(\omega_1) + \neg\mathrm{CH}$, *there are no Souslin trees.*

Proof Suppose that we are in a model of $MA(\omega_1) + \neg\mathrm{CH}$ and that we have a Souslin tree T. We recall that T is a partial order and we wish to turn this partial order into a forcing notion.

First, we lose nothing if we assume that the tree is rooted, so we can consider it as a forcing notion with the least element the root r, and moreover, by a standard combinatorial argument presented in [II,5.11 [63]], the tree T has a well pruned subtree. This means that it is a subtree with a root in which every element has an extension at every level. Any subtree of a Souslin tree is still Souslin, so we shall simply assume that T is well pruned.

Claim 12.1.4 T is a ccc forcing notion.

Proof (of Claim 12.1.4) Two elements p, q of T are incompatible in the forcing if they do not have a common extension in the tree, which will happen iff p and q are incompatible in the tree. So any antichain in the forcing with T is actually an antichain in T as a tree. Since T is a Souslin tree, it has no uncountable antichains, and therefore the forcing is ccc. $\bigstar_{12.1.4}$

Now we exhibit the dense sets needed. For any $\alpha > \omega_1$ we consider

$$\mathcal{D}_\alpha = \{p \in T : \mathrm{ht}(p) \geq \alpha\}.$$

By the assumption of T being well pruned, we have that each $\mathcal{D}_\alpha$ is dense.

Suppose now that G is a filter that intersects each set $\mathcal{D}_\alpha$. Since G is a filter, every two conditions in it are compatible, hence G is a chain in the tree. Since G intersects each $\mathcal{D}_\alpha$ it must be uncountable. Finally, this is a contradiction since T has no uncountable branches. $\bigstar_{12.1.3}$

Let us now consider if we can get more knowledge about Souslin trees by using $MA(\omega_1) + \neg$CH. For example, can we simply start with an Aronszajn tree and add to it an uncountable branch by forcing finite pieces of it? Perhaps in that way we shall get more information about the tree than just that it is not Souslin. Indeed, there is such a result, presented in Theorem 12.1.5 below. We start with a false try, to let the reader think a bit about the nature of forcing arguments, and to show the difficulty.

A false try.

Suppose that we are in a model of $MA(\omega_1)+\neg$CH and that T is an Aronszajn tree. Without loss of generality, by replacing T by an isomorphic copy if necessary, for every $\alpha < \omega_1$ the level α of T is enumerated by a subset of the ordinals in $[\omega\alpha, \omega\alpha + \omega)$. This implies that whenever $\gamma <_T \delta$, then $\gamma + \omega \leq \delta$. We shall show that T has an uncountable chain. By the argument on well pruned trees from [II,5.11 [63]], T has a well pruned subtree, as in the proof of Theorem 12.1.3. Such a subtree of T is then necessarily Aronszajn, so we can assume that T itself is well pruned. We shall add our branch by finite approximations, that is, let

$$\mathbb{P} = \{p : p \text{ a finite chain in } T\}$$

ordered by inclusion $p \leq q$ iff $p \subseteq q$. We note that $\mathbb{P}$ is a forcing notion with the least element $\emptyset$.

Now we need to exhibit dense sets which will show that $\mathbb{P}$ adds an uncountable chain to T. For $\alpha < \omega_1$, let

$$\mathcal{D}_\alpha = \{p \in \mathbb{P} : (\exists x \in p)\,\mathrm{ht}(x) \geq \alpha\}.$$

For each α, if p is not already a condition in $\mathcal{D}_\alpha$, denoting by $x = \max(p)$, we can use the assumption that T is well pruned to find $y >_T x$ such that $\mathrm{ht}(y) \geq \alpha$. Then letting $q \cup \{y\}$ gives an extension of p which is in $\mathcal{D}_\alpha$.

Applying $MA(\omega_1)$ to the family $\mathcal{F}$ of all sets $\mathcal{D}_\alpha$, we get an $\mathcal{F}$-generic filter G. It is clear that $\bigcup G$ gives an uncountable branch in T, as required.

It remains to prove that the forcing is ccc.

Let $\bar{p} = \langle p_\alpha : \alpha < \omega_1 \rangle$ be a sequence of conditions in $\mathbb{P}$ which we suppose, for a contradiction, forms an antichain. Therefore, each p_α is a finite set of ordinals in ω_1 which are pairwise compatible in the $<_T$-relation. Using the Δ-System Lemma, we can assume that p_αs form a Δ-System with root p^*. By throwing away countably many conditions if necessary, we can assume that $\max(p^*) < \min(p_\alpha \setminus p^*)$. We can similarly use a proof by induction to show that without loss of generality we may assume that $\max(p_\alpha \setminus p^*) < \min(p_\beta \setminus p^*)$ for all $\alpha < \beta < \omega_1$. For such $\alpha < \beta$, if any element x of $p_\alpha \setminus p^*$ is comparable in $<_T$ with any element y of $p_\beta \setminus p^*$, then $x <_T \min(p_\beta \setminus p^*)$, since the $<_T$-predecessors of y are well ordered and p_β is a chain in $<_T$, hence increasing as a set of ordinals in which the ordinal order and the $<_T$-order agree. Similarly, for all $z < x$ with $z \in p_\alpha$, we have $z <_T \min(p_\beta \setminus p^*)$. This analysis shows that the assumption that $\bar{p}$ is antichain, implies that for each α there is the first $n = n(\alpha)$ such that the n-th element in the increasing order of $p_\alpha \setminus p^*$ is not $<_T$ than $\min(p_{\alpha+1}) \setminus p^*$. By passing to an uncountable subset if necessary, we may assume that $n(\alpha)$ is a fixed value n. Let x_α be the n-th element of $p_\alpha \setminus p^*$.

Now we are tempted to state that $\langle x_\alpha : \alpha < \omega_1 \rangle$ must be an antichain in T, which is a contradiction with T being Aronszajn. However, there is no reason for this to be true, as we could have for example that $x_\alpha <_T x_{\alpha+2}$ for some α.

A more sophisticated version of the above proof, where one forces with finite pieces of a specialising function, works and gives a stronger result, due to James E. Baumgartner, Jerome Malitz and William N. Reinhardt in [7]:

Theorem 12.1.5 *Under* $MA(\omega_1) + \neg$CH, *every Aronszajn tree is special.*

An important observation used in the proof of Theorem 12.1.5 is the following Claim 12.1.6, here taken from Jech's book [51], where one can find a proof. In fact, although the book states the Claim in terms of Aronszajn trees, the same proof works for any tree of height and cardinality ω_1, as long as the tree does not have an uncountable branch.

Claim 12.1.6 ([51], Lemma 16.18) If **T** is a tree of height and cardinality ω_1 with no uncountable branches and W is an uncountable collection of finite pairwise disjoint subsets of **T**, then there exist $s, s' \in W$ such that any $x \in s$ is incomparable with any $y \in s'$.

After the discovery of Martin's Axiom a large number of applications of it were found, in topology, analysis and measure theory. The idea of being able to do forcing without entering into the logical difficulties proved to be attractive to mathematicians of diverse backgrounds. The contemporary state of the art of most of these applications is contained in David H. Fremlin's book [35]. About 50 years after its discovery now, it is rather rare to see new applications of Martin's Axiom, but there are still some. One is contained in our recent work with Shelah [26], where we answered an old question raised by Alan H. Mekler and Jouko Väänänen in [75]. We showed that under $MA(\omega_1) + \neg$CH there is no tree of size and height ω_1 and with no uncountable branches that weakly embeds all such trees.

13
Some More Large Cardinals

To develop more knowledge of forcing we have to go beyond the theory of ZFC and discuss large cardinals. We explore why this is the case and introduce some concrete large cardinal notions, including the supercompact cardinal. We shall, however, motivate that by going through a well known example of a large cardinal forcing first.

13.1 Prikry Forcing

We have seen that a forcing notion that preserves cofinalities automatically preserves cardinals, but is there a forcing that preserves cardinals and not cofinalities? Such a forcing was discovered by Karel Přikrý in [88]. We shall follow the common usage of the americanised version of Přikrý's name when discussing Prikry forcing.

The idea of this forcing is to take a measurable cardinal and to change its cofinality to ω, without changing any cardinals. Why measurable? Because the proof that cardinals are preserved is based on the so called Prikry Lemma, Lemma 13.1.4 below, which is proved using some combinatorial properties which imply a large cardinal strength of the underlying cardinal.

A measurable cardinal owes its name to its history of development through the 'Measure Problem', which asked if it was possible to extend the Riemann integral so that all real functions would be integrable. The question was asked by Henri Lebesgue in [68] and Lebesgue's measure is a partial answer to this. Vitali's discovery of a Lebesgue non-measurable subset of $\mathbb{R}$ in the presence of the Axiom of Choice [124] rejuvenated the question. Through the work of several researchers and various generalisations and simplifications (see the book [55] by Kanamori for a detailed history), the question finally settled to asking for the existence of a cardinal which is now called *a measurable cardinal* and which was defined by Stanisław Ulam in [120] as follows.

Definition 13.1.1 An uncountable cardinal κ is *measurable* if there is a non-principal ultrafilter $\mathcal{U}$ on κ which is closed under intersections of $< \kappa$-many elements. The latter property is called *κ-completeness.*

We recall that a filter $\mathcal{F}$ is non-principal if $\bigcap \mathcal{F} = \emptyset$. Being a measurable cardinal is a large cardinal property, since it was proved by Ulam [120] (also credited to Alfred Tarski and Ulam) that

Theorem 13.1.2 *A measurable cardinal is strongly inaccessible.*

Compare this with Theorem 7.2.5. In fact the property of being a measurable cardinal is much stronger than just being strongly inaccessible. For one, it can be proved that we may assume that for any measurable cardinal κ there is a κ-complete ultrafilter $\mathcal{U}$ on κ which is in addition *normal.* This means that for every sequence $\langle A_\alpha : \alpha < \kappa \rangle$ of elements in $\mathcal{U}$ we have

$$\Delta_{\alpha<\kappa} A_\alpha = \{\beta < \kappa : (\forall \alpha < \kappa)\beta \in A_\alpha\} \in \mathcal{U}.$$

Such an ultrafilter is called *a normal measure.* Secondly, a normal measure satisfies a strong partition property, as discovered by Frederick Rowbottom [94]: for every function $f : [\kappa]^{<\omega} \to \{0, 1\}$, there is a set $A \in \mathcal{U}$ such that for all $n < \omega$, the restriction $f \upharpoonright [A]^n$ is constant. (For details, see [55].) Now we are ready to define Prikry forcing.

Definition 13.1.3 Suppose that κ is a measurable cardinal and $\mathcal{U}$ is a κ-complete normal ultrafilter on κ. *Prikry forcing over $\mathcal{U}$*, denoted by $\Pr(\mathcal{U})$ consists of pairs (s, F) where s is a finite subset of κ and $F \in \mathcal{U}$.

The order is given by $(s, F) \le (t, H)$ if s is an initial segment of t, $H \subseteq F$ and $t \setminus s \subseteq F$.

The idea of this forcing is that the generic sequence $s^* = \bigcup\{s : (\exists F)(s, F) \in G\}$ determines an ω-sequence that is cofinal in κ. Since any two conditions of the form $(s, F), (s, H)$ are compatible, and there are only κ many finite subsets of κ, it follows that $\Pr(\mathcal{U})$ has κ^+-cc and preserves cardinals $> \kappa$. However, for preserving cardinals $\le \kappa$, we need to use the measurability of κ, as we shall now see.

Lemma 13.1.4 (Prikry Lemma) *Suppose that φ is a sentence of the forcing language of* $\Pr(\mathcal{U})$ *(so a sentence involving names in* $\Pr(\mathcal{U})$*, which will have truth value 0 or 1 in the extension) and that $(s, F) \in$* $\Pr(\mathcal{U})$*. Then there is an extension (s, H) of (s, F) which forces a truth value to φ.*

The point of the Prikry Lemma is that the extension (s, H) is obtained from (s, F) without changing the first coordinate. We shall see the use of that in Corollary 13.1.5. Such extensions are called *pure extensions* and we denote the situation by $(s, F) \leq^* (s, H)$. The idea of pure extensions goes back to Prikry's Ph.D. adviser Jack H. Silver at Berkeley and Silver forcing; one can see this in [73].

Proof For a finite subset t of $F \setminus (\max(s) + 1)$, define $f(t) = 1$ if there is X such that $(s \cup t, X)$ forces that φ is true, and let $f(t) = 0$ otherwise. Let $H \in \mathcal{U}$ be such that $f \restriction [H]^n$ is constant for every n. In particular $H \subseteq \kappa \setminus (\max(s) + 1)$. Note that by definition $H \subseteq F$. We claim that (s, H) forces a truth value to φ.

If not, there are extensions (s_l, H_l) for $l < 2$ of (s, H) such that each (s_l, H_l) forces the truth value l to φ. By extending further if necessary, we may assume that s_0 and s_1 are of the same length, say n, and therefore $t_0 = s_0 \setminus s$ and $t_1 = s_1 \setminus s$ obtain the same value under f. Yet $f(t_l) = l$, contradicting the choice of H. $\bigstar_{13.1.4}$

Corollary 13.1.5 *Prikry forcing preserves cardinals $\leq \kappa$.*

Proof First notice that if we manage to prove that all cardinals $< \kappa$ are preserved, then κ is preserved as well, since being inaccessible, κ is a limit cardinal and so it remains a cardinal if all cardinals below it remain cardinals. Suppose that $\lambda < \theta < \kappa$ is such that there is an injection $f : \theta \to \lambda$ in the Prikry extension. Let $\underset{\sim}{f}$ be a name for f and $p = (s, F)$ a condition that forces $\underset{\sim}{f}$ to be an injection. For every $\alpha < \theta$ and $\zeta < \lambda$ let $\varphi_{\alpha,\zeta}$ be the forcing sentence $\underset{\sim}{f}(\alpha) = \zeta$. Given $\alpha < \theta$, by the Prikry Lemma, we can find $H^\alpha_\zeta \subseteq F$ $(\zeta < \lambda)$ such that (s, H^α_ζ) decides the truth value of $\varphi_{\alpha,\zeta}$. Let $H_\alpha = \bigcap_{\zeta<\lambda} H^\alpha_\zeta$, hence (s, H_α) is an extension of (s, F) that forces a value to $\underset{\sim}{f}(\alpha)$. Now let $H = \bigcap_{\alpha<\theta} H_\alpha$. Then we have that $\underset{\sim}{f}(\alpha) = \zeta$ iff $(s, H) \Vdash \varphi_{\alpha,\zeta}$, which is definable in the ground model. Therefore $\underset{\sim}{f}$ is a function from the ground model and so θ and λ must have the same cardinality to start with. $\bigstar_{13.1.5}$

Generalisations of Prikry forcing abound, but they are sometimes quite difficult technically. An important generalisation to large cardinal forcing was discovered by Magidor in [70] which he used to prove the consistency of the failure of the Singular Cardinal Hypothesis modulo large cardinals. Much more about that subject will be said in Chapter 17. A reference article for Prikry-type forcing is Moti Gitik's [42]. A generalisation in another direction, where the measurable ultrafilter is replaced by an ultrafilter on ω and the generic object is a subset of ω, was discovered by Adrian R.D. Mathias in [73] and has since become a cornerstone of the Set Theory of the Reals. These and other generalisations of Prikry forcing to add new subsets of ω can be found in Thomas Jech's book [50].

13.2 Supercompact Cardinals and Laver Preservation

We have seen in Section 13.1 how useful it was to have a large cardinal to sustain the combinatorial properties of Prikry forcing. The large cardinal we used was measurable. As useful as it has proved in constructing Prikry forcing, a measurable cardinal is not very useful in arguments involving iterated forcing, because it is not very easy to preserve this large cardinal property when forcing with a single step forcing, let alone an iteration. See Observation 17.2.1 for an example of a forcing notion which destroys measurability.

There are many large cardinal notions that strengthen that of a measurability. The easiest way to define them is to reconsider the measurable cardinal from the logical point of view. Suppose that κ is a measurable cardinal and V is a model of ZFC. The assumption that κ is measurable implies the existence of a model of ZFC, by inaccessibility. Let $\mathcal{U}$ be a normal measure on κ and consider the *ultrapower* of V modulo $\mathcal{U}$, denoted by $M = \Pi V/\mathcal{U}$. Namely, the elements of M are the equivalence classes of $\Pi_\kappa V$ (the product of κ copies of V) under the relation $\sim$ defined by

$$f \sim g \iff \{\alpha < \kappa : f(\alpha) = g(\alpha)\} \in \mathcal{U}.$$

We interpret the relation $\in$ in M as $f \in g \iff \{\alpha < \kappa : f(\alpha) \in g(\alpha)\} \in \mathcal{U}$ and then we can prove that M is a model of ZFC. The completeness of the filter is used to obtain the well foundedness of the $\in^M$-relation.

Now, as in any ultrapower construction, we have an elementary embedding $j : V \to M$ given by assigning to any $x \in V$ the equivalence class of the identity x function. It is then rather easy to see that $j(\alpha) = \alpha$ for any $\alpha < \kappa$, but by considering the identity function [id] we can conclude that $j(\kappa) > \kappa$. We say that κ is the *critical point of j* and we write $\kappa = crit(j)$. Similar arguments show that M is closed under sequences of length $< \kappa$ and that $j(\kappa) < \kappa^{++}$.

The above presentation is not completely precise, in particular in that we should be using the so called 'Scott trick' for taking the ultrapowers of a proper class, invented by Scott in [97] in order to obtain that M is an inner model of V. A detailed presentation of this entire subject, along with a historical perspective is in [55], §5.

To conclude, from a measurable cardinal we get an elementary embedding. The opposite is true as well, as expressed by the following theorem from Keisler and Tarski [57].

Theorem 13.2.1 *A cardinal κ is measurable iff there exists an elementary embedding $j: V \to M$ with crit$(j) = \kappa$.*

Theorem 13.2.1 gives rise to many possible strengthenings, by requiring stronger and stronger properties of j and/or of M. This is in the heart of the linear order between many large cardinal notions, giving a rich theory, the classical parts of which are described in detail in [55]. We shall content ourselves with one of the generalisations of a measurable cardinal, which is the notion of a *supercompact cardinal.*

Definition 13.2.2 For any $\lambda \geq \kappa$, a cardinal κ is said to be *λ-supercompact* if there is an elementary embedding $j: V \to M$ with $crit(j) = \kappa$ such that M is a transitive inner model with ${}^{\lambda}M \subseteq M$ and $j(\kappa) > \lambda$. A cardinal is *supercompact* iff it is *λ-supercompact* for all λ.

In fact, λ-supercompact elementary embeddings can be assumed to come from a certain measure called *λ-supercompact measure.* Such measures are ultrafilters on the set $[\lambda]^{\kappa}$ and they give rise to ultrapower embeddings similarly as in the case of measurable cardinals. The main reason that supercompact cardinals are interesting for the theory of iterated forcing is the following preservation theorem by Laver [66].

Theorem 13.2.3 *If κ is supercompact, then there is a κ-cc forcing notion Q with $|Q| = \kappa$ such that in the extension by Q, κ is supercompact and remains supercompact upon forcing with any $(< \kappa)$-directed closed forcing.*

In this theorem we have used the notion of a *$(< \kappa)$-directed closed forcing,* which means for a forcing notion $\mathbb{P}$, that every subset A of $\mathbb{P}$ which is of size $< \kappa$ and *directed,* in the sense that every two elements have a common upper bound, has a common upper bound in $\mathbb{P}$. We shall see below in the proof of the consistency of PFA 15.5.2 how Theorem 13.2.3 gets to be used in conjunction with iteration of forcing. In fact Theorem 13.2.3 rests upon an important lemma, which introduces a $\diamond$-like notion called *Laver diamond.*

Definition 13.2.4 If κ is a supercompact cardinal, then there is a function $f^*: \kappa \to V_{\kappa}$ satisfying that for every set x and a cardinal $\lambda \geq |\mathrm{TC}(x)| + \kappa$, there is a supercompact measure $\mathcal{U}$ on $[\lambda]^{\kappa}$ such that for the elementary embedding j induced by $\mathcal{U}$ (so crit$(j) = \kappa$), we have

$$j(f^*)(\kappa) = x.$$

Here TC(x) stands for the *transitive closure of x*, which is the set obtained from x by closing under the operation '$y \in x$'. We shall see a use of Laver diamond in the consistency proof of PFA, Theorem 15.5.2. To motivate PFA and other Forcing Axioms stronger than MA, we first investigate the limitations of MA that have naturally led to the introduction of these axioms, in Chapter 14.

14

Limitations of Martin's Axiom and Countable Supports

A very well known early instance that shows a limitation of Martin's Axiom is Richard Laver's proof of the consistency of the Borel Conjecture [65]. Laver proved that this problem, which seems along the lines of those that have been solved using Martin's Axiom, on cardinal invariants of the continuum, *cannot* be answered using Martin's Axiom. A subset X of $\mathbb{R}$ is said to have *strong measure zero* if for every sequence $(\varepsilon_n)_n$ of positive real numbers, there is a sequence of intervals $(I_n)_n$ such that $\lg(I_n) < \varepsilon_n$ and $X \subseteq \bigcup_{n<\omega} I_n$. It is clear that all countable subsets of $\mathbb{R}$ have strong measure 0 and the Borel Conjecture from [8] states that the only strong measure zero sets are the countable ones.

In his paper [65], Laver first proves that Martin's Axiom does not provide a model in which the Borel Conjecture is true (Theorem 1.6 in [65]):

Theorem 14.0.1 *MA implies that there is a strong measure zero set of size* $2^{\aleph_0}$.

Is it possible then to obtain a model in which the Borel Conjecture is true? Laver [65] goes on to invent two things: Laver forcing with Laver trees and countable support iteration to iterate them. The idea here is that one starts with a ground model satisfying GCH and then does an iteration in which every step adds a function $r : \omega \to \omega$ whose presence assures that all ground model strong measure zero sets are countable. This is so, because whenever $X \subseteq [0, 1]$ is uncountable, there exists a final segment of r which codes a sequence $\langle \varepsilon_n : n < \omega \rangle$ such that X cannot be covered by the union $\bigcup_{n<\omega} I_n$ of intervals such that $\lg(I_n) < \varepsilon_n$. We shall explain that forcing now.

14.1 Laver Reals and Countable Support Iterations

Conditions in Laver forcing are subtrees T of ${}^{\omega>}\omega$ such that:

- T has a *stem* s_T, which is a node such that $(\forall s \in T) s \leq_T s_T \vee s_T \leq_T s$,
- for all $s \in T$ such that $s_T \leq_T s$, the set $\{n < \omega : s \frown \langle n \rangle \in T\}$ is infinite.

These conditions, called *Laver trees*, are ordered by reverse inclusion, so that T extends S, written as $S \leq T$ iff $T \subseteq S$. The latter of itemised properties is referred to as *infinite branching*.

A *Laver real* r is obtained from the generic filter by letting $r = \bigcup_{T \in G} s_T$. The size of ${}^{\omega>}\omega$ is $\aleph_0$ so the size of Laver's forcing is $2^{\aleph_0}$ and hence it satisfies $(2^{\aleph_0})^+$-cc. If we start with a model V of GCH, this means that the forcing satisfies $\aleph_2$-cc and hence preserves the cardinals $\geq \aleph_2$. However, consider an almost disjoint family $\{A_\alpha : \alpha < 2^{\aleph_0}\}$ of subsets of ω, that is a family of infinite subsets of ω such that any two have a finite intersection (such a family exists by a classical theorem discovered independently by Wacław Sierpiński, Hausdorff, Grigorii Fichtengolz–Leonid Kantorovich–see [II 1.3 [63]]). Now constructing conditions p_α for every $\alpha < 2^{\aleph_0}$ where for any $s \geq s_{p_\alpha}$ and any $n < \omega$, we have $s \in p_\alpha$ iff $n \in A_\alpha$, we obtain an antichain of size $2^{\aleph_0}$ in Laver forcing. Therefore, Laver forcing is not ccc and to see that it preserves ω_1, one has to use a different technique, that of *Fusion*.

Fusion is an elaboration of the pure extensions used by Prikry in that we define a sequence $\langle \leq_n : n < \omega \rangle$ of partial orders as follows.

First observe that is possible to order ${}^{\omega>}\omega$ in order type ω so that $s < t$ if $s \subseteq t$ and $s \frown \langle n \rangle < s \frown \langle n+1 \rangle$ are satisfied for all $s, t \in {}^{\omega>}\omega$ and $n < \omega$. This order induces an order on $\{s \in T : s_T \leq s\}$, for every Laver tree T, in which the least condition is the stem s_T. Let us enumerate $\{s \in T : s_T \leq s\}$ increasingly in $<$ as $s_T = s_0^T, \ldots s_n^T, \ldots$ For Laver trees p, q and $n < \omega$ we define:

$$p \leq_n q \iff p \leq q \text{ and for all } i \leq n, s_i^q = s_i^p.$$

It is clear that $\langle \leq_n : n < \omega \rangle$ form a refining sequence of partial orders on the set of Laver trees. They give the forcing a similar sort of auxiliary closure that we have seen in Prikry forcing with pure extensions, made precise by the following Lemma 14.1.1.

Lemma 14.1.1 *Suppose that $\langle p_n : n < \omega \rangle$ is a sequence of Laver trees such that $p_n \leq_n p_{n+1}$ holds for all n. Then $p = \bigcap_{n<\omega} p_n$ is a Laver tree satisfying $p_n \leq p$ for all n.*

Proof Notice that the requirements imply that all p_n have the same stem, s_0^0, and that $s_n^{p_m} = s_n^{p_n}$ for all $m \geq n$. Therefore $p = \{s_n^{p_n} : n < \omega\}$, p has the stem s_0^0 and it also easily seen that p is infinitely branching above the stem. $\bigstar_{14.1.1}$

The analogue of the Prikry Lemma in this situation is the following Lemma

14.1.2, which we give without proof. In it we use the notation $\veebar$ for the exclusive $\vee$.

Lemma 14.1.2 *Suppose that $m < \omega$ and φ_k $(k < m)$ are sentences of the forcing language in Laver forcing such that some condition $p \Vdash \veebar_{k<m}\varphi_k$. Then for each $n < \omega$ there is $p^n \geq_n p$ and $A_n \subseteq m$ with $|A_n| \leq n+1$ such that $p_n \Vdash \veebar_{k\in A_n}\varphi_k$.*

Using Lemma 14.1.2 we can prove that Laver forcing preserves ω_1, as follows.

Lemma 14.1.3 *Suppose that p is a condition in Laver forcing and*

$$p \Vdash (\underset{\sim}{f} : \omega \to \omega_1^V).$$

Then there is $q \geq_0 p$ and a countable $A \subseteq \omega_1^V$ such that $q \Vdash (\mathrm{ran}(\underset{\sim}{f}) \subseteq A)$.

Proof Starting with $p_0 = p$, we shall construct a sequence $\langle p_n : n < \omega\rangle$ of Laver trees such that $p_n \leq_n p_{n+1}$ holds for all n and countable sets $\langle A_n : n < \omega\rangle$ such that the fusion p^* of $\langle p_n : n < \omega\rangle$ satisfies $p^* \Vdash \mathrm{ran}(\underset{\sim}{f}) \subseteq \bigcup_{n<\omega} A_n$.

We will need the notion of a *fusion into* or *amalgamation*. Suppose that p is a Laver tree and t is a $\subseteq$-maximal node of $\{s_0^p, \dots s_n^p\}$, while $q \geq p$ is a condition with stem t. A fusion of q into p is given by $r = q \cup \{s \in p : s \perp t\}$. It can be seen that r is a Laver tree and that $r \geq_n p$. We can similarly define a simultaneous amalgamation into p of several conditions $q_0 \dots q_m$ whose stems are respectively all maximal nodes $\{t_0, \dots t_m\}$ among $\{s_0^p, \dots s_n^p\}$.

Let $\langle j_n : n < \omega\rangle$ be an enumeration of ω in which every $n < \omega$ appears infinitely often. Given p_n, consider $\{s_0^{p_n}, \dots s_n^{p_n}\}$ and let us enumerate all $\subseteq$-maximal nodes of $\{s_0^{p_n}, \dots s_n^{p_n}\}$ as $\{t_0, \dots t_k\}$, so $k \leq n$. For each $i \leq k$, if there exists $q_i \geq p_k$ with stem t_i and such that $q_i \Vdash \underset{\sim}{f}(j_n) = \alpha_n^i$, we choose some such $q_i = q_i^n$ and α_n^i. Let $A_n = \{\alpha_n^i : i \leq k, \alpha_n^i \text{ defined}\}$. Let p_{n+1} be the condition obtained from p_n by the fusion of all q_i into p_n, so $p_{n+1} \geq_n p_n$. Let p^* be the fusion of $\langle p_n : n < \omega\rangle$. We claim that $p^* \Vdash \mathrm{ran}(\underset{\sim}{f}) \subseteq \bigcup_{n<\omega} A_n$.

To prove this, suppose that $q \geq p^*$ and let $j < \omega$. Then there is $q' \geq q$ and some $\alpha < \omega_1$ such that $q' \Vdash \underset{\sim}{f}(j) = \alpha$. Let n be large enough such that $j_n = j$ and that the stem of q' is among $\{s_0^{p_n}, \dots s_n^{p_n}\}$. Therefore there must be a node $t \in q'$ which is maximal among $\{s_0^{p_n}, \dots s_n^{p_n}\}$ (as $q' \geq p_n$). Let us say that t was enumerated as t_i and the n-stage of the induction that constructed $\langle p_n : n < \omega\rangle$. Because of the existence of q', we have chosen and fused a q_i^n into p_n. Let $r = q' \cap \{s : s \subseteq t \vee t \subseteq s\}$, hence $r \geq q', q_i$ and hence $r \Vdash$ '$\underset{\sim}{f}(j) = \alpha_n^i$', which is an element of A. ★14.1.3

There is nothing special about ω_1 in Lemma 14.1.3, the same lemma holds with any set from V in place of ω_1. We were particularly interested in ω_1 because of the following immediate corollary:

Corollary 14.1.4 *Laver forcing preserves* ω_1.

The technique of fusion applies already to Sacks reals discovered by Gerald Sacks in [96], where he forced with perfect closed sets (or perfect trees) in order to obtain a real of minimal Turing degree. However, Laver's work on the Borel Conjecture brought another new element to the story of forcing: iterations with countable support. The definition is the same as that of iterations with finite supports in Definition 12.0.3, but we change the last item accordingly. We give the entire definition for completeness.

Definition 14.1.5 A *countable support iteration* of forcing of length α^* is defined by induction on α^* as a sequence of the form

$$\langle P_\alpha, \underset{\sim}{Q}_\beta : \alpha \leq \alpha^*, \beta < \alpha^* \rangle,$$

satisfying:

- $P_0 = \{\emptyset\}$ is the trivial forcing notion,
- for every $\alpha < \alpha^*$, $\underset{\sim}{Q}_\alpha$ is a P_α-name of a forcing notion,
- P_α is a forcing notion with the order denoted as $\leq_\alpha$,
- $P_{\alpha+1} = P_\alpha * \underset{\sim}{Q}_\alpha$, for every $\alpha < \alpha^*$,
- for every limit $\delta \in (0, \alpha^*]$,

$$P_\delta = \{p_\delta = (p_i : i < \delta) : (\forall i < \delta) p_\delta \restriction i \in P_i \text{ and } \mathrm{supt}(p_\delta) \text{ is countable}\},$$

is a forcing notion ordered by $p_\delta \leq_\delta q_\delta$ iff for every $i < \delta$, $p_\delta \restriction i \leq_i q_\delta \restriction i$.

Laver proved that a countable iteration of Laver forcing preserves ω_1. This is done using a version of the fusion argument from Lemma 14.1.3 now given in the following form, proved as Lemma 5 in [65].

Lemma 14.1.6 *Suppose that P_{α^*} is the result of a countable support iteration of Laver forcing, $\langle F_n :, n < \omega \rangle$ is an increasing sequence of finite subsets of α^* and $\langle p_n : n < \omega \rangle$ is a sequence of conditions of P_{α^*} such that for every n, $p_n \leq p_{n+1}$ and*

$$(\forall \alpha \in F_n) p_{n+1} \restriction \alpha \Vdash_{P_\alpha} \text{`} p_n(\alpha) \leq_n^\alpha p_{n+1}(\alpha)\text{'},$$

where $\leq_n^\alpha$ denotes the derived order $\leq_n$ with respect to $\underset{\sim}{Q}_\alpha$ as forced by P_α (the above situation is denoted by $p_n \leq_n^{F_n} p_{n+1}$). Further assume that $\bigcup_{n<\omega} F_n = \bigcup_{n<\omega} \mathrm{supt}(p_n)$. Then there is a $p_\omega \in P_{\alpha^*}$ such that for all n we have $p_n \leq_n^{F_n} p_\omega$.

It remains to see that the Laver real does the coding it was constructed for and this again is proved using a fusion lemma. This finishes the review of Laver's forcing.

Based on the ideas on forcing with trees in Sacks and Laver forcing, Baumgartner [5] isolated a property which guarantees the preservation of ω_1 and which is preserved by iterations with countable supports, called Axiom A. He proved that this property is satisfied by many forcing notions that add reals, including Mathias, Laver and Sacks reals and that every ccc forcing satisfies it (Lemma 14.1.8 below). Thanks to this property Baumgartner gave another proof of the consistency of the Borel Conjecture, using Mathias forcing. The preservation of the Axiom A property under countable support iterations is proved similarly to the iteration property of Laver forcing.

Definition 14.1.7 A forcing notion $(\mathbb{P}, \leq)$ satisfies *Axiom A* if there are orders $\langle \leq_n : n < \omega \rangle$ on $\mathbb{P}$ such that:

- $\leq_0 = \leq$ and $p \leq_{n+1} q \implies p \leq_n q$,
- if $\langle p_n : n < \omega \rangle$ is a sequence of condition such that for every n, $p_n \leq_n p_{n+1}$, then there is $q \in \mathbb{P}$ such that $q \geq_n p_n$ for all n,
- if $A \subseteq \mathbb{P}$ is an antichain, then for each $p \in \mathbb{P}$ and $n < \omega$, there is $q \geq_n p$ such that $\{r \in A : r \text{ is compatible with } q\}$ is countable.

It is clear that every countably closed forcing satisfies Axiom A, as one can just let $\leq_n = \leq$ for all n. We give a proof that also every ccc forcing satisfies Axiom A.

Lemma 14.1.8 *Every ccc forcing satisfies Axiom A.*

Proof Suppose that $\mathbb{P}$ is a ccc forcing notion. Let $\leq_0 = \leq$ and for $n \geq 1$ let $\leq_n$ be the equality. Then the first two items of the definition of Axiom A are clearly satisfied and the last one is obtained by taking $q = p$. $\bigstar_{14.1.8}$

15
Proper Forcing and PFA

A new chapter in the theory of forcing was opened by Shelah's discovery of proper forcing in [100]. The novelty of Shelah's approach is a move from infinite combinatorics to the technique of elementary submodels, which has since integrated set theory to the extent that it is hard to imagine a forcing argument that does not use it. A major reference on the technique and results about proper forcing is Shelah's book [105]. Throughout this chapter, we fix an uncountable regular cardinal χ which is so large that all arguments relative to the forcing notions we discuss are contained in the set $\mathcal{H}(\chi)$ of all sets x with $|\mathrm{TC}(x)| < \chi$. We equip $\mathcal{H}(\chi)$ by some canonical well order $<^*$ and use the notation V to stand for the ground model where our forcing takes place. We shall be taking elementary submodels of the structure $(\mathcal{H}(\chi), \in, <^*)$, which we for simplicity again denote by $\mathcal{H}(\chi)$. To be able to define what proper forcing is, we need some background in $\mathcal{P}_\kappa\lambda$-combinatorics, which we define and discuss in the next section.

15.1 $\mathcal{P}_\kappa\lambda$-combinatorics

The combinatorial properties of clubs and stationary sets can be generalised from the context of ordinals to various other contexts, see Jech's article [52] for a survey. We shall be interested in a generalisation where the ordinals are replaced by 'small' sets of ordinals, as we now describe. The notions here are due to Jech in [53], who used the following notation.

Definition 15.1.1 For two infinite cardinals $\kappa < \lambda$, we denote by $\mathcal{P}_\kappa\lambda$ the family $\{A \subseteq \lambda : |A| < \kappa\}$.

Of course the same notion is also denoted by $[\lambda]^{<\kappa}$ in the context of cardinal partition relations, see [30] for example, but the above notation is standard when studying the object with respect to clubs and stationary sets and we have wanted the reader to know it. We shall only be interested in the situation when $\kappa = \aleph_1$, so we shall formulate our definitions in that way, although they apply more generally. We shall use $[\lambda]^{<\kappa}$ more often than we use $\mathcal{P}_\kappa \lambda$.

Definition 15.1.2 For an uncountable set A whose cardinality is of uncountable cofinality, a subset C of $[A]^{\leq\aleph_0}$ is called *closed* if it is closed under countable unions of its members. C is called *unbounded* if for every $x \in [A]^{\leq\aleph_0}$ there is $y \in C$ such that $x \subseteq y$. A *club* in $[A]^{\leq\aleph_0}$ is a closed and unbounded set.

A *stationary* subset of $[A]^{\leq\aleph_0}$ is one that intersects every club.

Notice that for every $W \subseteq [A]^{\leq\aleph_0}$ there is the smallest closed set C with $W \subseteq C$, which is the set obtained by taking all unions of countable $\subseteq$-increasing sequences of sets in W. Such a set is called the *closure* of W.

It is an exercise to prove the following lemma, which can also be found as Lemma 1.3. in [105].

Lemma 15.1.3 *The intersection of countably many clubs of* $[A]^{\leq\aleph_0}$ *is a club.*

Thanks to Lemma 15.1.3 we have that the club sets generate a countably closed filter on $[A]^{\leq\aleph_0}$, which is the filter of all sets that contain a club and which we denote by $\mathcal{D}_{\aleph_0}(A)$. It is important to note that in the context of $[\mathcal{H}(\chi)]^{\leq\aleph_0}$, it follows from the Löwenheim–Skolem theorem that for every element $a \in \mathcal{H}(\chi)$, the set of all countable $M \prec \mathcal{H}(\chi)$ with $a \in M$ forms a club of $[\mathcal{H}(\chi)]^{\leq\aleph_0}$.

15.2 Definition and Properties of Proper Forcing

The following is a main definition in this context.

Definition 15.2.1 (1) Suppose that $M \prec \mathcal{H}(\chi)$ and $\mathbb{P}$ is a forcing notion in M. Say that a subset $\mathcal{D} \subseteq \mathbb{P}$ is *open* if for all $p \in \mathcal{D}$ and $q \geq p$, we have $q \in \mathcal{D}$.

A condition $q \in \mathbb{P}$ is $(\mathbb{P}, M)$-*generic* if for every open dense subset $\mathcal{D}$ of $\mathbb{P}$ which happens to be in M, the set $\mathcal{D} \cap M$ is *predense* above q, that is

$$r \geq q \implies (\exists s \in \mathcal{D} \cap M)\,(r \text{ and } s \text{ are compatible}).$$

(2) A forcing notion $\mathbb{P}$ is *proper* if for every $p \in \mathbb{P}$ and a parameter $a \in \mathcal{H}(\chi)$, there is a club set of countable models $M \prec \mathcal{H}(\chi)$ with $a, p, \mathbb{P} \in M$ and such there is a $(\mathbb{P}, M)$-generic condition $q \geq p$.

The point of a $(\mathbb{P}, M)$-generic condition for a countable M is that above such a condition the forcing is ccc, relative to the antichains that are in M. (Indeed, the reader might like to try to reformulate the definition of a generic condition in terms of antichains in order to get convinced by the previous sentence.) In fact, there is a strong analogy between the idea of a generic condition and a much earlier notion of a *master condition* by Silver in the early 1970s; see more about this after the proof of the consistency of PFA, Theorem 15.5.2. There are several equivalent characterisations of genericity and properness, some involving combinatorics and others involving games. One that is quite practical is the following seeming strengthening of the definition of properness (see Theorem 2.8 of [105]).

Theorem 15.2.2 *A forcing notion $\mathbb{P}$ is proper if for every $p \in \mathbb{P}$, for every countable $M \prec \mathcal{H}(\chi)$ with $p, \mathbb{P} \in M$, there is a $(\mathbb{P}, M)$-generic condition $q \geq p$.*

Another interesting characterisation is given by the following lemma.

Lemma 15.2.3 *Suppose that $M \prec \mathcal{H}(\chi)$ is countable and $\mathbb{P}$ is a forcing notion in M. Then a condition q is $(\mathbb{P}, M)$-generic iff for every $\mathbb{P}$-name $\underset{\sim}{\tau} \in M$ of an object in V, q forces the value of $\underset{\sim}{\tau}$ to be in $M[\underset{\sim}{G}]$.*

Proof Suppose first that q is $(\mathbb{P}, M)$-generic. Let

$$\mathcal{D} = \{r \in \mathbb{P} : r \Vdash \text{ a value for } \underset{\sim}{\tau}\}.$$

Since $\mathcal{D}$ is defined using $\mathbb{P}$ and $\underset{\sim}{\tau}$, which are both in M and M is elementary, hence closed under definability, we have that $\mathcal{D} \in M$. Being a $\mathbb{P}$-name means that above every condition there is one which forces a value to the name, hence $\mathcal{D}$ is dense. Therefore, by properness $\mathcal{D} \cap M$ is predense above q. Note that every $s \in \mathcal{D} \cap M$ forces a value for $\underset{\sim}{\tau}$ which is in $M[\underset{\sim}{G}]$, again by elementarity. Therefore every condition above q can be extended to one that forces a value for $\underset{\sim}{\tau}$ which is in $M[\underset{\sim}{G}]$, which finishes the proof.

For the other direction, suppose that $\mathcal{D} \in M$ is an open dense subset of $\mathbb{P}$ and that $r \geq p$, we need to find a condition in $\mathcal{D} \cap M$ which is compatible with r. We shall produce a name $\underset{\sim}{\tau}$ for an ordinal in M, as follows. In M we can enumerate $\mathcal{D}$ as $\mathcal{D} = \{s_i : i < i^*\}$ for some ordinal $i*$ (although this does not imply that all $s_i \in M$). If G a is $\mathbb{P}$-generic filter which contains q, then $m = \min\{i < i^* : s_i \in G\}$ is in $M[G]$ by the fact that the enumeration is in M and that q forces the names in M to be evaluated in $M[G]$. Let $\underset{\sim}{m}$ be a name for m in M, therefore $q \Vdash (\exists i \in M)(\underset{\sim}{m} = s_i)$. Since $r \geq q$, r forces the same thing. This means that in any generic H which contains r, there must be an element of $\mathcal{D} \cap M$, and therefore r is compatible with such an element. ★15.2.3

Lemma 15.2.3 implies that for a proper forcing, we have $M[G] \cap V = M$. One should also note the following general theorem, see Theorem 2.11 in [105].

Lemma 15.2.4 *Suppose that $M \prec \mathcal{H}(\chi)$ and $\mathbb{P}$ is a forcing notion in M, while G is $\mathbb{P}$-generic over V. Then in $V[G]$ we have that $M[G] \prec \mathcal{H}(\chi)^{V[G]}$.*

15.3 Axiom A and Proper Forcing

Properness is a more general notion than Axiom A. One can find a game-theoretic argument for this in Baumgartner [6]. Let us give an argument based on the notions developed above.

Theorem 15.3.1 *Every Axiom A forcing notion is proper.*

We shall need the following observation.

Observation 15.3.2 *Suppose that $M \prec H(\chi)$ and $B \in M$ is countable. Then $B \subseteq M$.*

Proof Since $\omega, B \in M$, the fact that there is a surjection from ω to B is verified by M and hence M has such a surjection f. For every $n < \omega$ we have $f(n) \in M$, and hence $B = \{f(n) : n < \omega\}$ is a subset of M. $\bigstar_{15.3.2}$

Proof of Theorem 15.3.1 Suppose that $\mathbb{P}$ is a forcing notion that satisfies Axiom A, as witnessed by the orders $\langle \leq_n : n < \omega \rangle$, that $p \in \mathbb{P}$ and that a parameter $a \in \mathcal{H}(\chi)$ is given. Let $M \prec \mathcal{H}(\chi)$ be countable and such that $a, p, \mathbb{P}, \langle \leq_n : n < \omega \rangle \in M$. We shall construct a countable $M^* \supseteq M$ such that there is a $(\mathbb{P}, M^*)$-generic $q \geq p$.

Let $\langle A_n : n < \omega \rangle$ be the list of maximal antichains in $\mathbb{P}$ that are elements of M. By induction on $n < \omega$, we choose $\langle M_n : n < \omega \rangle$ and $\langle p_n : n < \omega \rangle$ such that:

- $p_0 = p$ and $M_0 = M$,
- $p_n \leq_n p_{n+1}$,
- $M_n \prec H(\chi)$ is countable,
- for every n, letting $\{A^n_m : n < m < \omega\}$ enumerate all maximal antichains of $\mathbb{P}$ that are elements of M_n, we have that for every $m > n$, the set

$$B^n_m \stackrel{\text{def}}{=} \{r \in A^n_m : r \text{ is compatible with } p_{n+1}\}$$

  is countable,
- $\langle B^n_m : m < \omega \rangle, p_{n+1}, M_n \in M_{n+1}$.

To do the induction, once we have constructed p_n and M_n, we notice that since M_n is countable, there is indeed an enumeration $\{A^n_m : n < m < \omega\}$ as required.

Now we use Axiom A and induction on n to choose $\langle q^n_m : n < m < \omega\rangle$ such that $q^n_{n+1} = p_n$, $q^n_m \leq_m q^n_{m+1}$ and such that the set

$$\{r \in A^n_m : r \text{ is compatible with } q^n_{m+1}\}$$

is countable. We let p_{n+1} be the fusion of $\langle q^n_m : n < m < \omega\rangle$. Hence $p_n \leq_n p_{n+1}$ and since $p_{n+1} \geq q^n_{m+1}$, then the set $B^n_m \subseteq \{r \in A^n_m : r \text{ is compatible with } q^n_{m+1}\}$ is countable. Having defined p_{n+1} we can choose M_{n+1} by an application of the Löwenheim–Skolem theorem.

At the end of the induction we let $M^* = \bigcup_{n<\omega} M_n$ and let q be the fusion of $\langle p_n : n < \omega\rangle$. Clearly, $M^* \prec H(\chi)$ is countable. We claim that q is $(\mathbb{P}, M^*)$-generic. To see this, suppose that $t \geq q$ and that $\mathcal{D}$ is an open dense subset of $\mathbb{P}$ with $\mathcal{D} \in M^*$. Let $\mathcal{A}$ be a maximal antichain contained in $\mathcal{D}$ and such that $\mathcal{A} \in M^*$, which exists by elementarity. Let n such that $\mathcal{A} \in M_n$. By density of $\mathcal{D}$, the antichain $\mathcal{A}$ is actually maximal in $\mathbb{P}$, not just maximal in $\mathcal{D}$. Therefore $\mathcal{A} = A^n_m$ for some $m < \omega$. We have that $t \geq q \geq p_{n+1}$, so the set $\{r \in A^n_m : r \text{ is compatible with } t\} \subseteq B^n_m \subseteq M_{n+1} \subseteq M^*$. Therefore there is $r \in \mathcal{A} \cap M^*$ which is compatible with t, as required.

Analysing the above argument, it clearly shows that for every $n < \omega$

$$W_n \stackrel{\text{def}}{=} \{N \text{ countable} : N \prec \mathcal{H}(\chi), p, \mathbb{P} \in N \text{ and there is a } (\mathbb{P}, N)\text{-generic } q \geq_n p\}$$

forms an unbounded subset of $[\mathcal{H}(\chi)]^{\leq\aleph_0}$. Combining this with a fusion argument, we obtain that the set of all countable $N \prec \mathcal{H}(\chi)$ with $p, \mathbb{P} \in N$ and such that there is $q \geq p$ which is $(\mathbb{P}, N)$-generic, is a club. Since p is arbitrary, this shows that $\mathbb{P}$ is proper. $\bigstar_{15.3.1}$

An example of a proper forcing which does not satisfy Axiom A is Baumgartner's forcing for adding a club to ω_1 using finite conditions (see §3 in [6]).

15.4 Iteration of Proper Forcing with Countable Supports

The heart of the classical arguments about proper forcing rests upon a beautiful diagonalisation argument of Shelah, who proved the following theorem.

Theorem 15.4.1 *Suppose that $\langle P_\alpha, \underset{\sim}{Q}_\beta : \alpha \leq \alpha^*, \beta < \alpha^*\rangle$ is a countable support iteration in which for every $\alpha < \alpha^*$,*

$$\Vdash_{P_\alpha} \text{'}\underset{\sim}{Q}_\alpha \text{ is proper'}.$$

Then P_{α^*} is proper.

The rest of this section will be devoted to the proof. Before presenting the proof itself we shall present some facts about iteration of forcing which will be useful in the proof.

For any iterated notion of forcing

$$\langle P_\alpha, \underset{\sim}{Q}_\beta : \alpha \le \alpha^*, \beta < \alpha^* \rangle,$$

we introduce the following notation.

Notation 15.4.2 Suppose that $\alpha < \alpha^*$, $q \in P_\alpha$ and $q \ge_\alpha p \restriction \alpha$. We define $q \uplus p$ as the condition $q \frown p \restriction [\alpha, \alpha^*)$, noting that this is a well defined condition and that $q \uplus p \ge_{\alpha^*} p$.

For simplicity in the presentation of the main proof, we shall ignore the cases of distinct conditions $p_0 \le p_1$ and $p_1 \le p_0$, which can perfectly appear in the case of iterated forcing. We shall simply treat this situation as $p_0 = p_1$, as such conditions are forcing equivalent, that is, one of them is in the generic filter if the other one is. However, this presentational simplification means that several other points of the proof are oversimplified. Therefore we suggest to the reader to read the presentation here first, to get the most important ideas, and then to read the presentation in [105, Theorem 3.2].

Proof (outline) The proof is by induction on α^*, for all forcing notions simultaneously. The trick of the induction is to make an inductive hypothesis which is somewhat stronger than the final desired conclusion. It is:

Inductive Hypothesis Suppose that $M \prec \mathcal{H}(\chi)$ is countable and such that $\langle P_\alpha, \underset{\sim}{Q}_\beta : \alpha \le \alpha^*, \beta < \alpha^* \rangle \in M$. Then, for all $p \in P_{\alpha^*} \cap M$, for all $\alpha \in \alpha^* \cap M$, *if* $q \in P_\alpha$ is (P_α, M)-generic and $q \ge p \restriction \alpha$, then there is a (P_{α^*}, M)-generic r such that $r \ge p$ and $r \restriction \alpha = q$.

Now we do the induction. The case $\alpha^* = 0$ is clear since P_0 is the trivial forcing notion which is clearly proper.

Suppose that $\alpha^* = \alpha+1$, so $P_{\alpha^*} = P_\alpha * \underset{\sim}{Q}_\alpha$. We have $\alpha^* \in M$, so $\alpha \in M$. Since the inductive hypothesis holds up to α, it suffices to consider the instance of it between α and α^*. Let q be as in the assumptions of the inductive hypothesis and let G_α be P_α-generic with $q \in G_\alpha$. Suppose $p = (p_0, \underset{\sim}{p}_1) \in M \cap (P_\alpha * \underset{\sim}{Q}_\alpha)$.

By the assumptions, we have that $\underset{\sim}{Q}_\alpha[G_\alpha]$ is a proper forcing and that $p_1 \in Q[G_\alpha]$. By the properness of P_α and the fact that $\underset{\sim}{p}_1 \in M$, it follows that $p_1 \in M[G_\alpha]$ (see Lemma 15.2.3). By Lemma 15.2.4, we have that $M[G_\alpha] \prec \mathcal{H}(\chi)[G_\alpha] = \mathcal{H}(\chi)^{V[G_\alpha]}$. Since $Q_\alpha[G_\alpha]$ is a proper forcing, there is $s \ge_{Q_\alpha[G_\alpha]} p_1$ which is $(M[G_\alpha], Q_\alpha[G_\alpha])$-generic. Since all that was needed of the generic here is that $q \in G_\alpha$, we have that q forces the existence of s. Therefore, by the

Existential Completeness Lemma 8.4.7 there is a P_α-name $\underset{\sim}{s}$ such that

$$q \Vdash `\underset{\sim}{s} \geq_{\underset{\sim}{Q}_\alpha} p_1 \text{ is } (M[\underset{\sim}{G}_\alpha], \underset{\sim}{Q}_\alpha)\text{-generic'}. \tag{15.1}$$

We let $r = q \frown (\alpha, s)$. Then clearly $r \restriction \alpha = q$ and $r \geq p$. We have to prove that r is (M, P_{α^*})-generic. Suppose that $r' \in P_{\alpha^*}$ and $r' \geq r$.

Let $\mathcal{D} \in M$ be a dense open subset of P_{α^*}. Then the set $\mathcal{E} = \{t \restriction \alpha : t \in \mathcal{D}\}$ is a dense open subset of P_α which is in M, and therefore since $r' \restriction \alpha \geq q$, there is $q' \in \mathcal{E} \cap M$ with $q' \geq r' \restriction \alpha$. Let us choose in M a P_α-name $\underset{\sim}{O}$ for

$$\{\underset{\sim}{o} \in \underset{\sim}{Q}_\alpha : (\exists x \geq_{P_\alpha} q')(x, \underset{\sim}{o}) \in \mathcal{D}\}.$$

Hence $\underset{\sim}{O}$ is forced to be an open set in $\underset{\sim}{Q}_\alpha$ which is dense above $(q', \emptyset_\alpha)$ and to belong to $M[\underset{\sim}{G}_\alpha]$. Since $q' \geq q$ and $q' \Vdash r'(\alpha) \geq s$, we have by (15.1) that q' forces the existence of $\underset{\sim}{o} \in \underset{\sim}{O} \cap M[\underset{\sim}{G}_\alpha]$ such that $\underset{\sim}{o} \geq r'(\alpha)$. By elementarity of M we can assume that $\underset{\sim}{o} \in M$. Let x be a witness that $\underset{\sim}{o} \in \underset{\sim}{O}$. Again by elementarity, we can assume that $x \in M$ and then $(x, \underset{\sim}{o}) \in \mathcal{D} \cap M$ is above r'.

Now we go to the case of α^* being a limit ordinal. Let us fix $\alpha < \alpha^*$ and we show how to preserve the induction hypothesis. Let q be as in the assumptions, for a given $p \in P_{\alpha^*} \cap M$. Let $\{\mathcal{D}_n : n < \omega\}$ be an enumeration of open dense subsets of P_{α^*} that are in M. Since $\alpha, \alpha^* \in M$, there is a sequence $\langle \alpha_n : n < \omega \rangle$ in M such that $\sup_{n<\omega} \alpha_n = \alpha^*$ and such that $\alpha_0 = \alpha$. By induction on n we define two sequences $\langle p_n : n < \omega \rangle$ and $\langle q_n : n < \omega \rangle$ so that:

- $p_n \in P_{\alpha^*} \cap M$ and $p_0 = p$,
- $q_n \in P_{\alpha_n}$ is (M, P_{α_n})-generic,
- $q_n \geq_{\alpha_n} p_n \restriction \alpha_n$,
- $q_{n+1} \restriction \alpha_n = q_n$ (see a remark in the footnote)[1]
- $p_{n+1} \restriction \alpha_n = p_n \restriction \alpha_n$,
- $q_n \uplus p_{n+1} \in \mathcal{D}_n$.

To do the induction, it is possible to choose q_0 by the induction hypothesis, since it implies that P_{α_n} is proper. Suppose now that p_n, q_n are given. First we need the following claim, whose proof is left to the reader.

Claim 15.4.3

$$\mathcal{E}_n = \{r \restriction \alpha_n : r \in \mathcal{D}_n \text{ and } r \geq_{\alpha^*} p_n\}$$

is open dense in P_{α_n}.

[1] At a closer study of this proof, the reader might see this and the following requirement as a Shelah's diagonalisation version of the order $\leq_n$ in Laver's or Baumgartner's countable iteration argument for fusion.

Now note that $\mathcal{E}_n$ belongs to M. Since q_n is (M, P_{α_n})-generic, we have that there is $q^+ \in \mathcal{E}_n \cap M$ with $q^+ \geq q_n$. By the definition of $\mathcal{E}_n$, there is $r \in \mathcal{D}_n$ such that $r \restriction \alpha_n = q^+$ and $r \geq_{\alpha^*} p_n$. By elementarity, we can find such r in M. Let $p_{n+1} = p_n \restriction \alpha_n \uplus r \restriction [\alpha_n, \alpha^*)$.

To define q_{n+1} we shall apply the inductive hypothesis between α_n and α_{n+1} so that $q_{n+1} \restriction \alpha_n = q_n$, the condition q_{n+1} is $(M, P_{\alpha_{n+1}})$-generic and $q_{n+1} \geq p_{n+1}$ (doing this in complete detail requires a bit of argument).

Coming to the end of the induction, the point is that $\langle q_n : n < \omega \rangle$ has a least upper bound $q^* = \bigcup_{n<\omega} q_n$. Clearly $q^* \geq p$ and $q^* \restriction \alpha = q$. We have to show that q^* is (M, P_{α^*})-generic. First we remark that $q^* \geq p_n$ for all n. Now suppose that $r^* \geq_{\alpha^*} q^*$ and that $\mathcal{D}$ is open dense in P_{α^*} with $\mathcal{D} \in M$. Then $\mathcal{D}$ was enumerated as $\mathcal{D}_n$ for some n. By the choice of q_{n+1}, we have that q^* and hence r^* is above an element of $\mathcal{D}_n \cap M$, namely p_{n+1}. $\bigstar_{15.4.1}$

One may wonder if countable supports are necessary for the iteration of proper forcing, that is, if it is not the case that finite support is sufficient for all applications. The following Example 15.4.4 from Martin Goldstern's paper [47, 2.0.2] shows an inherent limitation of finite supports.

Example 15.4.4 Suppose that $\langle P_\alpha, \underset{\sim}{Q}_\beta : \alpha \leq \alpha^*, \beta < \alpha^* \rangle$ is a finite support iteration of forcing in which every step has at least two incompatible conditions, and that $\alpha^* \geq \omega$. Then $V[G_{\alpha^*}]$ contains a real which is Cohen generic over V. Namely, taking simply $\alpha^* = \omega$, let q_n^l for $l < 2$ and $n < \omega$ be such that $\Vdash_n q_n^0 \perp_{\underset{\sim}{Q}_n} q_n^1$. Define $f \in V[G_\omega]$ by letting $f(n) = l$ iff $q_n^l \in G_n$. To see that f is Cohen over V, taking any $g \in {}^\omega 2 \cap V$ and $n < \omega$, we have that the set $\{q \in P_\omega : (\exists m \geq n) q(m) = q_m^{1-g(l)}\}$ is dense, because every condition has a finite support and hence we can find a large enough m where we have not yet decided the value of the m-th coordinate and can declare it to differ from g.

A typical example of a forcing which does not add Cohen reals is Sacks forcing, which also does not add Cohen reals in an $\aleph_2$-iteration by countable support, say over **L**. One can see more on this for example in the paper by Stefan Geschke and Sandra Quickert [39].

15.5 Proper Forcing Axiom

Having seen that proper forcing is preserved by iterations with countable supports, it is very natural to try to imitate the proof of the consistency of MA to obtain a consistency proof for an axiom that applies to proper forcings, which will be called PFA, *proper forcing axiom*. So, instead of iterating ccc forcing with finite supports, we iterate proper forcing with countable supports. Let us formulate the axiom corresponding to MA(ω_1) in this context.

Definition 15.5.1 PFA is the axiom stating that for every proper forcing $\mathbb{P}$ and every family $\mathcal{F}$ of $\aleph_1$ dense subsets of $\mathbb{P}$, there is a filter G in $\mathbb{P}$ which intersects all sets in $\mathcal{F}$.

Some changes necessary to be able to adapt the proof of MA to the proof of PFA, become clear rather quickly. The first one is the question of the length of iteration. To obtain the consistency of MA(ω_1), for example, we start with a model of GCH and iterate all ccc forcings of size $\aleph_1$ in an iteration of length ω_2. This suffices *because* the property of ccc is reflexive to subsets of size $\aleph_1$: if every subset of size $\aleph_1$ of a given forcing notion $\mathbb{P}$ is ccc, then so is $\mathbb{P}$. This argument can be pushed a bit further to see that

MA(ω_1) holds iff MA(ω_1) holds when restricted to posets of size $\aleph_1$.

Using this property, we first do the bookkeeping argument as in Theorem 12.0.5 for ccc posets of size $\aleph_1$, and then we argue that the final model in the iteration for the consistency of MA(ω_1) satisfies not just MA(ω_1) restricted to posets of size $\aleph_1$, but also MA(ω_1). A similar reflection is not implied by the property of being proper, and indeed there is no reason to believe that a forcing notion in which every subset of size $\aleph_1$ is proper, is proper itself. A quite convincing way of seeing this is to consider the large cardinal strength, as follows.

To obtain PFA for proper forcing notions of size $\aleph_1$, we imitate the proof of the consistency of MA(ω_1) and we do not need to use any large cardinals. On the other hand, full PFA provably has large cardinal strength. For example, in a well known work Todorčević [117] proves that PFA implies the failure of the $\square_\kappa$ (see Definition 7.8.8) for all κ uncountable, which is known to have considerable large cardinal strength. We present Todorčević's proof in §16.4. In fact Solovay in [113] had proved that $\square_\kappa$ fails everywhere above a strongly compact cardinal, so Todorčević's result shows that in the presence of PFA, $\aleph_2$ has supercompactness properties. There have been more results of this type since then; let us mention Matteo Viale's proof in [123] that PFA implies the *Singular Cardinal Hypothesis* (SCH) (see Chapter 17 for more on SCH), while it was known from the work of Solovay in [113] that SCH holds above a strongly compact (so above a supercompact) cardinal. The work of Viale and Christoph Weiß in [126] shows that basically any consistency proof of PFA using a similar iteration scheme as the one by Baumgartner, will need to use a supercompact cardinal. A new proof of the consistency of PFA was given more recently by Neeman, as will be described in §16.5, but that proof also uses a supercompact cardinal. Viale and Weiß [126] show that under PFA, the cardinal $\aleph_2$ satisfies a combinatorial principle ISP which has the property that

$$\mathrm{ISP}(\kappa) + \kappa \text{ strongly inaccessible} \implies \kappa \text{ is supercompact.}$$

Hence, $\aleph_2$ under PFA is like a supercompact cardinal stripped off its inaccessibility. These results give a strong heuristic evidence for the conjecture that the consistency strength of PFA is exactly a supercompact cardinal.

Theorem 15.5.2 (Baumgartner) *From the consistency of the existence of a supercompact cardinal, one can conclude that it is consistent to have a model of PFA.*

Proof (outline) Suppose that κ is a supercompact cardinal and let f^* be a Laver diamond on κ (see 13.2.4). By induction on $\alpha \leq \kappa$ we define the forcing notion P_α and a P_α-name for a forcing notion $\underset{\sim}{Q}_\alpha$ for $\alpha < \kappa$, starting with P_0 being the trivial forcing. The iteration is made with countable supports.

Given P_α, if α is a limit ordinal and $f^*(\alpha) = (\underset{\sim}{Q}, \bar{\underset{\sim}{\mathcal{D}}})$ where $\underset{\sim}{Q}$ is a P_α-name of a proper forcing notion and $\bar{\underset{\sim}{\mathcal{D}}}$ is a P_α-name of a sequence of $\aleph_1$-many dense sets in $\underset{\sim}{Q}$, then we let $\underset{\sim}{Q}_\alpha = \underset{\sim}{Q}$. Otherwise, we let $\underset{\sim}{Q}_\alpha$ be a P_α-name for the collapse $\mathrm{Col}(\aleph_1, \aleph_2)$.

It follows by the preservation of properness by countable support iterations that P_κ is proper (notice that by being countably closed, $\mathrm{Col}(\aleph_1, \aleph_2)$ is proper), and hence it preserves $\aleph_1$. It is also easy to see that P_κ is κ-cc and hence preserves cardinals $\geq \kappa$. By the fact that we have cofinally often collapsed cardinals to $\aleph_1$, it follows that in the extension κ becomes $\aleph_2$.

Suppose that $\underset{\sim}{Q}$ is a P_κ-name for a proper forcing and $\bar{\underset{\sim}{\mathcal{D}}} = \{\underset{\sim}{\mathcal{D}}_i : i < \omega_1\}$ are names for a family of dense subsets of $\underset{\sim}{Q}$. Let λ be large enough so that all these names are in V_{λ_0} for some λ_0 with $2^{\lambda_0} < \lambda$, and we also assume that $2^{|P_\kappa * \underset{\sim}{Q}|} < \lambda$. We can without loss of generality assume that P_κ is an order whose domain is a subset of λ. Let $j : V \to M$ be a λ-supercompactness embedding with $\mathrm{crit}(j) = \kappa$ and such that $j(f^*)(\kappa) = (\underset{\sim}{Q}, \bar{\underset{\sim}{\mathcal{D}}})$ and let G be P_κ-generic over the ground model V. We shall use a technique of 'lifting the embedding', that is, we shall define an extension of j to a function which is an embedding in $V[G]$. First note that the whole forcing P_κ is contained in V_κ, so that $j \restriction P_\kappa = \mathrm{id}$. By elementarity, $j(\langle P_\alpha, \underset{\sim}{Q}_\beta : \alpha < \kappa, \beta < \kappa \rangle)$ is a forcing iteration of length $j(\kappa)$ and it is defined in the same way as $\langle P_\alpha, \underset{\sim}{Q}_\beta : \alpha < \kappa, \beta < \kappa \rangle$, but using $j(f^*)$ in place of f^*. Therefore the composition $\langle P_\alpha, \underset{\sim}{Q}_\beta : \alpha < \kappa, \beta < \kappa \rangle * \underset{\sim}{Q}$ forms an initial segment of $j(P_\kappa) = j(\langle P_\alpha, \underset{\sim}{Q}_\beta : \alpha < \kappa, \beta < \kappa \rangle)$. Let us denote the remaining final segment by $\mathbb{R}$. By these facts and the choice of G, we have that G is also P_κ-generic over M and therefore by the general facts about iterated forcing (see (12.1)) there are a Q-generic H_0 over $M[G]$ and a $\mathbb{R}$-generic H_1

over $M[G][H_0]$ such that $M[G][H_0][H_1]$ is a $j(P_\kappa)$-generic extension of M. We denote the $j(P_\kappa)$-generic corresponding to $M[G][H_0][H_1]$ by $G * H_0 * H_1$.

Claim 15.5.3 There is an extension j^* of j to $V[G]$ which is an elementary embedding of $V[G]$ into $M[G][H_0][H_1]$.

Proof There is a standard way to define j^*, due to Silver, but one has to prove that j^* is elementary. Every object in $V[G]$ is of the form $\underset{\sim}{\tau}_G$ for some P_κ name τ. Since P_κ is an initial segment of $j(P_\kappa)$, we can understand τ as a $j(P_\kappa)$-name (the reader might want to check this as an exercise). Therefore τ has an interpretation in $M[G][H_0][H_1]$, which we denote by $\underset{\sim}{\tau}_{G*H_0*H_1}$ and which we define to be $j^*(\underset{\sim}{\tau}_G)$.

To prove that j^* is elementary, we consider an arbitrary formula φ such that $V[G] \models \varphi(\underset{\sim}{\tau}_{0G}, \ldots \underset{\sim}{\tau}_{nG})$. This means that there is $p \in G$ with $p \Vdash_{P_\kappa} \varphi(\underset{\sim}{\tau}_0, \ldots \underset{\sim}{\tau}_n)$. But we have that $j(p) = p$ is an element of $G * H_0 * H_1$, so $M[G][H_0][H_1] \models \varphi(\underset{\sim}{\tau}_{0G*H_0*H_1}, \ldots \underset{\sim}{\tau}_{nG*H_0*H_1})$. Since the same argument works for $\neg\varphi$, we have proven the elementarity. $\bigstar_{15.5.3}$

It is customary to denote j^* again by j. So far our proof has been detailed, but now we shall skip a few details (see pages 105–106 in Keith Devlin's article [21]), to describe the main idea as the following. By the fact that j is a λ-supercompactness embedding, it follows that ${}^\lambda M \subseteq M$ and hence $j \restriction \lambda \in M \subseteq M[G][H_0]$. Again by the choice of λ it follows that H_0 is not only $\underset{\sim}{Q}$-generic over $M[G]$ but also over $V[G]$. Clearly H_0 intersects all sets in $\underset{\sim}{\bar{D}}_G$. Therefore in $M[G][H_0][H_1]$ we have a filter which intersects all sets in $\underset{\sim}{\bar{D}}$ and now we use the elementarity between $V[G]$ and $M[G][H_0][H_1]$ to conclude that such a filter must also exist in $V[G]$. $\bigstar_{15.5.2}$

The first printed version of the consistency proof of PFA is in Devlin's article [21]. It does not credit the proof to Baumgartner or to anybody else, neither does Baumgartner directly credit the proof to himself, even in his well known article [6] on applications of PFA, which may indicate a parallel development between him and Shelah. However, it is a common agreement that this proof is actually due to Baumgartner, see for example Kanamori's historical article [56]. Shelah's book [105] contains proofs of various versions of PFA. A well known further development of these ideas is in the article [34] by Matthew Foreman, Magidor and Shelah, where they prove that a forcing axiom SPFA previously considered by Shelah is in fact equivalent to a maximal axiom that one can obtain along these lines, the *Martin Maximum* MM.

Silver's idea of master conditions and lifting the embeddings was developed in the context of his proof that GCH can fail at a measurable cardinal. This work forms a basis of modern arguments but it was never published. It is usually referred to as [108], see for example Foreman and W. Hugh Woodin's article [33]. A modern reference on the connection between forcing and elementary embeddings is Cummings' article [14].

15.6 On $\aleph_1$-dense Sets of Reals

We have already mentioned some applications of PFA and the fact that many applications of proper forcing appear in Shelah's book [105]. Many others are in Baumgartner's article [6]. One of the best known combinatorial applications is due to Baumgartner, who showed that under PFA every two $\aleph_1$-*dense sets of reals* are order isomorphic. Since that application has had influence on further developments covered in this book, we shall explain Baumgartner's theorem here.

Definition 15.6.1 Suppose that L is a linear order and that κ is a cardinal. Then L is κ-*dense* if for every $a < b$ in L, the interval (a, b) has size at least κ.

A known example of an $\aleph_0$-dense countable linear order is the set $\mathbb{Q}$ of rationals. It is a theorem of Cantor that up to isomorphism, $\mathbb{Q}$ is the only such order which in addition has no end-points. Hausdorff [49] studied generalisations of this theorem and proved that under the assumption of $2^{<\kappa} = \kappa$ there is an order type η_κ which is κ-dense and, moreover, has the property that for every $A, B \subseteq \eta_\kappa$ both of size $< \kappa$ satisfying that every element of A is less than every element of B (written as $A < B$), there is c such that $A < c < B$. Such an order type is unique up to isomorphism, in fact it is an example of what in model theory is called a saturated model (for more details about Hausdorff's proof see Joseph Rosenstein's book [92]). As we shall see in Section 15.7, PFA is not consistent with CH, so it is interesting to ask if nevertheless one can have some version of uniqueness up to isomorphism for 'saturated' linear orders of size $\aleph_1$ under PFA. The following theorem was proved by Baumgartner in [4].

Theorem 15.6.2 (Baumgartner) *PFA implies that every two $\aleph_1$-dense sets of reals are isomorphic.*

In fact, the partial order used for the proof of Theorem 15.6.2 is a composition of a countably closed forcing followed by a ccc forcing. Therefore, like many applications of PFA, the theorem does need the full PFA. The theorem gives rise to several questions. The first one is if the countably closed component is necessary, that is if one could obtain the same result just with ccc forcing. This question was answered in the negative by Uri Abraham (Avraham) and Shelah in [2].

Theorem 15.6.3 (Abraham and Shelah) *MA does not imply that all $\aleph_1$-dense sets of reals are isomorphic.*

The second question is to ask if PFA is simply an axiom for forcing of the type countably closed forcing followed by a ccc forcing. This question is also answered in the negative, as we discuss in Section 15.7. Next, we may wonder if an analogue of Theorem 15.6.2 might be consistent for $\aleph_2$-dense sets of reals under the assumption of $2^{\aleph_0} \geq \aleph_3$. Much partial progress on this question was made by Neeman, likely leading to the answer in the affirmative; see more discussion and references in §16.5.

15.7 Further on PFA

In this section we briefly discuss forcing axioms of the type:

FA(Γ): for every forcing notion $\mathbb{P}$ of type Γ and every family $\mathcal{D}$ of $\aleph_1$ dense sets in $\mathbb{P}$, there is a filter H in $\mathbb{P}$ which intersects all sets in $\mathcal{D}$.

In the model obtained in the proof of Theorem 15.5.2, the value of the continuum is $2^{\aleph_0} = \aleph_2$. Given that MA is consistent with arbitrary large values of the continuum, it is natural to ask if PFA is as well. A joint result of Boban Veličković and Todorčević from the 1980s, presented in Todorčević's book [118] is that PFA actually implies that $2^{\aleph_0} = \aleph_2$. Veličković in [121] showed that the analogue of PFA where we only consider countably closed followed by ccc forcing, already implies $2^{\aleph_1} = \aleph_2$.

Theorem 15.7.1 (Veličković) *Let Γ consist of all forcing notions that are of the type countably closed followed by ccc forcing. Then FA(Γ) implies $2^{\aleph_1} = \aleph_2$.*

The consistency strength of the forcing axiom from Theorem 15.7.1 is exactly that of one weakly compact cardinal. In conjunction with Todorčević's result from [117], this shows that PFA is strictly stronger than the corresponding axiom for countably closed followed by ccc forcing, since one weakly compact cardinal is not enough to fail $\square_\kappa$ for all uncountable κ. The fact that PFA implies $2^{\aleph_0} = \aleph_2$ has been improved in a different direction by Justin T. Moore in [82]. He proved that a bounded version of PFA called BPFA already implies $2^{\aleph_0} = \aleph_2$. BPFA has the strength of one reflecting cardinal and is sufficient to carry the proof of Theorem 15.6.2. Moore's paper also introduces a combinatorial principle MRP (Mapping Reflection Principle) which is implied by PFA and which in some sense codes the difference between PFA and

FA(countably closed * ccc). In particular, it is MRP which is responsible for the fact that PFA implies SCH in the work of Viale [123].

16

$\aleph_2$ and other Successors of Regulars

The iteration theorems we presented so far give a wealth of results about the subsets of ω and, to some extent, ω_1. For example, a typical result obtained by the application of iterated proper forcing is the construction of a model in which $\omega_1 = \square < \blacksquare = \omega_2$, where $\square$ and $\blacksquare$ are cardinal invariants of the continuum. One can consult the book [3] by Tomek Bartoszyński and Haim Judah for various instances of this method. The difficulties of generalising this context to the larger cardinals, even of the form κ^+ for κ satisfying $\kappa^{<\kappa} = \kappa$, turn out to be substantial. In the above mentioned exposition of iterated forcing Baumgartner [5, §4] wrote

> The search for extensions of MA for larger cardinals has proved to be rather difficult.

One reason for this is that for κ regular and uncountable, an iteration of $(< \kappa)$-closed and κ^+-cc forcing posets with supports of size less than κ does not in general have the κ^+-cc. For example, the following is a theorem due to Mitchell and described in a paper of Laver and Shelah [67].

Theorem 16.0.1 *Suppose, for simplicity*[1] *that* $\mathbf{V} = \mathbf{L}$. *Then there are countably closed* $\aleph_2$*-Souslin trees* $\langle T_n : n < \omega \rangle$ *such that the product of any finitely many of the trees still has* $\aleph_2$*-cc, but* $\otimes_{n<\omega} T_n$ *does not.*

The above shows that in $\mathbf{L}$ there is an iteration of length ω of countably closed $\aleph_2$-cc forcing such that the inverse limit at stage ω does not have the $\aleph_2$-cc.

The literature contains several preservation theorems for iterations involving strengthened forms of closure and chain condition, along with corresponding forcing axioms. The first results in this direction are in unpublished work by Laver (see [5, §4]). Baumgartner [5] proved that under CH an iteration with

[1] All we need is CH and $\diamond(S^2_1)$.

countable supports of countably compact $\aleph_1$-linked forcing posets is $\aleph_2$-cc, and proved the consistency of some related forcing axioms. Shelah [99] proved that under CH an iteration with countable supports of posets which are countably closed and well-met and which enjoy the $\aleph_2$-stationary cc (see Definition 16.1.1(2) below), also enjoys the $\aleph_2$-stationary cc. Shelah also proved more general results for certain iterations of κ^+-stationary cc posets with supports of size less than κ, and proved the consistency of a number of related forcing axioms. We do not define the notions needed for the main Baumgartner's and Shelah's iteration theorems since we shall present a theorem that generalises them both, Theorem 16.1.3 below. It is a result from[§1 [17]], due to Cummings, Džamonja, Magidor, Charles Morgan and Shelah.

16.1 A version of MA for Successors of Regulars

This section is based on [§1 [17]] by Cummings, Džamonja, Magidor, Morgan and Shelah. Let us start by a definition.

Definition 16.1.1 (1) Let $(\mathbb{P}, \leq)$ be a forcing notion.

- Two increasing sequences $(q_i)_{i<\omega}$ and $(r_i)_{i<\omega}$ from $\mathbb{P}$ are *pointwise compatible* if for each $i < \omega$ we have that q_i and r_i are compatible.
- $(\mathbb{P}, \leq)$ is *countably parallel-closed* if each pair of pointwise compatible increasing ω-sequences has a common upper bound.

(2) Suppose that κ is a regular cardinal. A forcing notion $\mathbb{P}$ is said to be κ^+-*stationary cc* if for every sequence $\langle p_i : i < \kappa^+\rangle$ of conditions in $\mathbb{P}$, there is a club subset C of κ^+ and a regressive function f such that

$$(\forall i, j < \kappa^+)[\mathrm{cf}(i) = \mathrm{cf}(j) = \kappa \text{ and } f(i) = f(j) \implies p_i, p_j \text{ are compatible}].$$

Clearly, κ^+-stationary cc forcing satisfies κ^+-cc. The following fact about κ^+-cc forcing will be useful.

Claim 16.1.2 Suppose that $\mathbb{P}$ is a κ^+-cc forcing and $\underset{\sim}{D}$ is a $\mathbb{P}$-name for a club of κ^+. Then there is a club C such that $\Vdash_{\mathbb{P}}$ '$C \subseteq \underset{\sim}{D}$'.

Proof Let $\{\underset{\sim}{d}_i : i < \kappa^+\}$ be a $\mathbb{P}$-name for an increasing enumeration of $\underset{\sim}{D}$. For every i, let $\mathcal{A}_i$ be a maximal antichain of conditions deciding the value of $\underset{\sim}{d}_i$. Let $\beta_i = \sup\{\alpha_i : (\exists p \in \mathcal{A}_i)\, p \Vdash$ '$\underset{\sim}{d}_i = \alpha_i$'$\}$. Notice that $\beta_i < \kappa^+$ since $|\mathcal{A}_i| \leq \kappa$. Then $\Vdash$ '$\underset{\sim}{d}_i \leq \beta_i$'. We can similarly show that for every $\alpha < \kappa^+$ there is $j_\alpha < \kappa^+$ such that $\Vdash$ '$\alpha < \underset{\sim}{d}_{j_\alpha}$'.

We define γ_i for $i < \kappa^+$ by induction on i. Let $\gamma_0 = \beta_0$. Given γ_i, let $\gamma_{i+1} =$

$\beta_{j_{\gamma_i}} + 1$ and for δ limit ordinal let $\gamma_\delta = \sup_{i<\delta} \gamma_i$. Notice that this definition has assured that $\Vdash$ '$(\gamma_i, \gamma_{i+1}) \cap \underset{\sim}{D} \neq \emptyset$', for all i. But then, since $\underset{\sim}{D}$ is forced to be a club, it is forced that $\gamma_\delta \in \underset{\sim}{D}$ for all limit δ and it suffices to let $C = \{\gamma_\delta : \delta \text{ limit } < \delta\}$. ★16.1.2

By an iteration with supports of size $< \kappa$ we mean exactly what the name suggests: every condition in the iteration has support of size $< \kappa$.

Theorem 16.1.3 (Cummings, Džamonja, Magidor, Morgan and Shelah) *Let κ be an uncountable regular cardinal with $\kappa^{<\kappa} = \kappa$. Every iteration of countably parallel-closed, $(< \kappa)$-closed, κ^+-stationary cc forcing with supports of size less than κ is $(< \kappa)$-closed and has the κ^+-stationary cc.*

In the proof, we shall use the notation $S^{\kappa^+}_\kappa = \{\alpha < \kappa^+ : \text{cf}(\alpha) = \kappa\}$.

Proof Let $\langle P_\alpha, \underset{\sim}{Q}_\beta : \alpha \leq \alpha^*, \beta < \alpha^* \rangle$ be a $(< \kappa)$-support iteration of forcings such that for each $\alpha < \alpha^*$ it is forced by P_α that $\underset{\sim}{Q}_\alpha$ is countably parallel-closed, $(< \kappa)$-closed and κ^+-stationary cc. We prove, by induction on α, that for each $\alpha \leq \alpha^*$, the forcing P_α has the stationary κ^+-chain condition and is $(< \kappa)$-closed.

Suppose that for all $\varepsilon < \alpha$ the forcing P_ε is $(< \kappa)$-closed and has the stationary κ^+-chain condition. It is then standard to conclude that P_α is $(< \kappa)$-closed, so let us concentrate on the stationary κ^+-chain condition. Let $\langle p_i : i < \kappa^+ \rangle \in {}^{\kappa^+} P_\alpha$. By induction on $n < \omega$, we construct for each $i < \kappa^+$ an increasing ω-sequence $\langle p_i^n : n < \omega \rangle$ such that $p_i^0 = p_i$, as follows.

Induction step $n + 1$. Suppose that we have already defined $\langle p_i^n : i < \kappa^+ \rangle$. For each $i < \kappa^+$ and $\varepsilon < \alpha$ we have $\Vdash_{P_\varepsilon}$ '$p_i^n(\varepsilon) \in \underset{\sim}{Q}_\varepsilon$ and $p_i^n(\varepsilon) = \emptyset_{q_\varepsilon}$ if $\varepsilon \notin \text{supt}(p_i)$'. As $\Vdash_{P_\varepsilon}$ '$\underset{\sim}{Q}_\varepsilon$ has the κ^+-stationary cc' and $\langle p_i^n(\varepsilon) : i < \kappa^+ \rangle$ names a κ^+-sequence of elements of $\underset{\sim}{Q}_\varepsilon$, we may find P_ε-names $\underset{\sim}{C}_\varepsilon^n$ for a club subset of κ^+ and g_ε^n for a regressive function on $\underset{\sim}{C}_\varepsilon^n$ witnessing the κ^+-stationary cc for the sequence named by $\langle p_i^n(\varepsilon) : i < \kappa^+ \rangle$. By Claim 16.1.2, since P_ε has the κ^+-chain condition, for every ε and n, there is a club D_ε^n such that

$$\Vdash_{P_\varepsilon} \text{`}D_\varepsilon^n \text{ is a club subset of } \underset{\sim}{C}_\varepsilon^n\text{'}.$$

So we may as well assume that $\underset{\sim}{C}_\varepsilon^n$ canonically names some club $C_\varepsilon^n \in V$.

For each $i < \kappa^+$, dealing with each $\varepsilon \in \text{supt}(p_i^n)$ inductively and using the $(< \kappa)$-closure of P_ε, we find some $p_i^{n+1} \geq p_i^n$ such that

$$(\forall \varepsilon \in \text{supt}(p_i^n))\, p_i^{n+1} \restriction \varepsilon \Vdash_{P_\varepsilon} \text{`}\underset{\sim}{g}_\varepsilon^n(i) = \rho_\varepsilon^n(i)\text{'}$$

for some ordinal $\rho_\varepsilon^n(i) < i$. Let $\rho_\varepsilon^n(i) = 0$ for ε which are not in $\text{supt}(p_i^n)$.

Let $\{\varepsilon_\alpha : \alpha < \mu\}$, for some $\mu \leq \kappa^+$, We note that

$$\bigcup \{\text{supt}(p_i^n) : n < \omega, i < \kappa^+\}$$

has cardinality $\leq \kappa^+$ and we let $\{\varepsilon_j : j < \mu\}$, for some cardinal $\mu \leq \kappa^+$ enumerate that set. For each $j < \mu$ let $C_{\varepsilon_j} = \bigcap_{n<\omega} C^n_{\varepsilon_j}$ and let

$$C = \{i < \kappa^+ : (\forall j < i)\, i \in C_{\varepsilon_j}\}.$$

Note that C is a club of κ^+.

The following is a version of the classical Δ-System Lemma that is needed here. Its proof uses the assumption $\kappa^{<\kappa} = \kappa$.

Claim 16.1.4 There is a a club $E \subseteq C$ and a regressive function g on $S^{\kappa^+}_\kappa \cap E$ such that if $i < i'$ are both in $S^{\kappa^+}_\kappa \cap E$ and $g(i) = g(i')$, then

(1) $\bigcup_{n<\omega} \operatorname{supt}(p^n_i) \cap \{\varepsilon_\gamma : \gamma < i\} = \bigcup_{n<\omega} \operatorname{supt}(p^n_{i'}) \cap \{\varepsilon_\gamma : \gamma < i'\}$,
(2) $\bigcup_{n<\omega} \operatorname{supt}(p^n_i) \subseteq \{\varepsilon_\gamma : \gamma < i'\}$,
(3) if $\gamma < i'$, $n < \omega$ and $\varepsilon_\gamma \in \operatorname{supt}(p^n_{i'})$, then $\rho^n_{\varepsilon_\gamma}(i) = \rho^n_{\varepsilon_\gamma}(i')$.

(Note that, by (1), for every γ to which clause (3) applies we have $\gamma < i$ and $\varepsilon_\gamma \in \bigcup_{n<\omega} \operatorname{supt}(p^n_i)$.)

Proof of 16.1.4 We start by making some auxiliary definitions.

Let $H = \{h : \operatorname{dom}(h) \in [\kappa^+]^{<\kappa}$ and $h : \operatorname{dom}(h) \longrightarrow {}^\omega\kappa^+\}$. By the hypothesis that $\kappa^{<\kappa} = \kappa$ we have that $|H| = \kappa^+$. So we can enumerate H as, say,

$$H = \{h_\eta : \eta < \kappa^+\}.$$

Noticing that $H \subset [\kappa^+ \times {}^\omega\kappa^+]^{<\kappa}$, for $i < \kappa^+$, define $H_i = H \cap [i \times {}^\omega i]^{<\kappa}$. Since $\kappa^{<\kappa} = \kappa$ we have, for each $i < \kappa$, that $|H_i| = \kappa$. So we can define $f : \kappa^+ \longrightarrow \kappa^+$ by, for $i < \kappa^+$, letting $f(i) =$ the least $\tau < \kappa^+$ such that $H_i \subseteq \{h_\eta : \eta < \tau\}$.

Also, for each $i < \kappa^+$, set $F_i = \{\gamma < \kappa^+ : \varepsilon_\gamma \in \bigcup\{\operatorname{supt}(p^n_i) : n < \omega\}\}$ and $D_i = F_i \cap i$.

Let $E = \{i < \kappa^+ : f``i \cup \bigcup\{F_j : j < i\} \subseteq i\}$. We claim that $E$ is a club of $\kappa^+$. In order to see this let us define for $n < \omega$ the functions $g_n : \kappa^+ \longrightarrow \kappa^+$ by $g_0(i)$ is the least ordinal $\zeta$ such that $f``i \cup \bigcup\{F_j : j < i\} \subseteq \zeta$, and $g_{n+1}(i) = g_0(g_n(i))$ for $n < \omega$. Then for any $i < \kappa^+$ we have that $i < \bigcup_{n<\omega} g_n(i) \in E$. This shows that E is unbounded. Moreover, if $B \subseteq E$ is of limit order type less than κ^+ and $d \in \sup(B) \cup \bigcup\{F_j : j < \sup(B)\}$ then there is some $b \in B$ such that $d \in b \cup \bigcup\{F_j : j < b\}$ and hence $f(d) < b < \sup(B)$.

Now define, for each $i < \kappa^+$, the function $t_i : D_i \longrightarrow {}^\omega i$ by letting

$$t_i(\gamma)(n) = \rho^n_{\varepsilon_\gamma}(i) \text{ if } \varepsilon_\gamma \in \operatorname{supt}(p^n_i) \text{ and } t_i(\gamma)(n) = 0 \text{ otherwise.}$$

So for each $i < \kappa^+$ we have $t_i \in H_i$. Define $g : \kappa^+ \longrightarrow \kappa^+$ by setting $t(i)$ to be that η such that $t_i = h_\eta$. For each $i < \kappa^+$ we have $t_i = h_{g(i)}$ and thus $g(i) < f(i)$.

Now if $i \in E$ and $\operatorname{cf}(i) = \kappa$, since t_i is a function of size $(< \kappa)$, there is some $i^* < i$ such that $t_i \in H_{i^*}$. Hence for such i we have $g(i) < f(i^*)$. We also have

$f(i^*) < i$ since $i \in E$. Therefore if $i \in E$ and $\mathrm{cf}(i) = \kappa$ we have $g(i) < i$ and so we have shown that g is regressive on $E \cap S^{\kappa^+}_\kappa$.

Moreover, if $i, i' \in E \cap S^{\kappa^+}_\kappa$ and $g(i) = g(i')$ we have that $t_i = t_{i'}$, and hence that (1) and (3) hold. Finally, if $i < i'$ as well we have, since $i' \in E$, that $F_i \subseteq i'$ and hence (2) holds. ★16.1.4

Let g be as provided by Claim 16.1.4 and suppose that $i < i' < \kappa^+$ are both in $S^{\kappa^+}_\kappa \cap E$ and that $g(i) = g(i')$. We construct, by an induction of length α, a condition $q \in P_\alpha$ which is a common refinement of p^n_i and $p^n_{i'}$ for all n, and hence of p_i and $p_{i'}$. The support of q will be $\bigcup_{n<\omega} \mathrm{supt}(p^n_i) \cup \bigcup_{n<\omega} \mathrm{supt}(p^n_{i'})$.

For $\sigma \in \bigcup_{n<\omega} \mathrm{supt}(p^n_i) \setminus \bigcup_{n<\omega} \mathrm{supt}(p^n_{i'})$, as $\Vdash_{P_\sigma}$ '$\underset{\sim}{Q}_\sigma$ is countably closed', take $q(\sigma)$ to be any r such that $q \restriction \sigma \Vdash_{P_\sigma} (\forall n < \omega)\, r \geq p^n_i(\sigma)$. Similarly, for $\sigma \in \bigcup_{n<\omega} \mathrm{supt}(p^n_{i'})$, take $q(\sigma)$ to be any r such that $q \restriction \sigma \Vdash_{P_\sigma} (\forall n < \omega)\, r \geq p^n_{i'}(\sigma)$.

Finally, if $\sigma \in \bigcup_{n<\omega} \mathrm{supt}(p^n_i) \cap \bigcup_{n<\omega} \mathrm{supt}(p^n_{i'})$ then $\sigma = \varepsilon_\gamma$ for some $\gamma < i$ (by conditions (1) and (2) above), for each $n < \omega$ we have $\rho^n_{\varepsilon_\gamma}(i) = \rho^n_{\varepsilon_\gamma}(i')$ (by (3) above), and $i, i' \in C_{\varepsilon_\gamma}$ (by the construction of C and the choice of E).

By construction, for each $n < \omega$ we have

$$q \restriction \sigma \Vdash_{P_\sigma} \text{`}p^n_i(\sigma) \text{ is compatible with } p^n_{i'}(\sigma)\text{'}.$$

By the fact that $\Vdash_{P_\sigma}$ '$\underset{\sim}{Q}_\sigma$ is countably parallel-closed', we can choose $q(\sigma)$ to be some r such that $q \restriction \sigma \Vdash_{P_\sigma} (\forall n < \omega)\, r \geq p^n_i(\sigma), p^n_{i'}(\sigma)$. Thus for all $n < \omega$ we have that $q \restriction (\sigma + 1) \geq p^n_i \restriction (\sigma + 1), p^n_{i'} \restriction (\sigma + 1)$. ★16.1.3

We remark that the previous theorem allows us to give 'generalised Martin's axiom' forcing axioms similar to those formulated by Baumgartner [5] and Shelah [99]. One example is given by the following theorem.

Theorem 16.1.5 *Let κ be an uncountable regular cardinal such that $\kappa^{<\kappa} = \kappa$, and let $\lambda > \kappa^+$ be a cardinal such that $\gamma^{<\kappa} < \lambda$ for every $\gamma < \lambda$. Then there is a $(< \kappa)$-closed and κ^+-stationary cc forcing poset $\mathbb{P}$ such that if G is $\mathbb{P}$-generic, then in $V[G]$ we have $2^\kappa = \lambda$ and the following forcing axiom holds:*

For every poset $\mathbb{Q}$ which is $(< \kappa)$-closed, countably parallel-closed and κ^+-stationary cc, every $\gamma < \lambda$ and every sequence $\langle \mathcal{D}_i : i < \gamma\rangle$ of dense subsets of $\mathbb{Q}$ there is a filter on $\mathbb{Q}$ which meets each set $\mathcal{D}_i$.

The exact connections between this forcing axiom and those given in [5] and [99] are explained in [17].

As an example of a forcing notion that satisfies the premises of Theorem 16.1.5, we claim that Cohen forcing does. Let us show that $\mathrm{Add}(\kappa, \lambda)$ is κ^+-stationary cc.

Suppose that $\{p^i : i < \kappa^+\}$ is a collection of conditions in $\mathrm{Add}(\kappa, \lambda)$. For each $i < \kappa^+$ let $\{\alpha_\gamma : \gamma < \gamma^*\}$ be an enumeration of $\bigcup_{i<\kappa^+} \mathrm{dom}(p_i)$, for some $\gamma^* \leq \kappa^+$. For each $i < \kappa^+$ let $\{\alpha^i_\gamma : \gamma < \gamma^i\}$, for some $\gamma^i < \kappa$, be the increasing enumeration of $\mathrm{dom}(p_i)$, let $\theta^i = \sup(\{\gamma : \alpha_\gamma \in \mathrm{dom}(p_i)\}) + 1$, let $T^i = \{\gamma < i : \alpha_\gamma \in \mathrm{dom}(p_i)\}$, and let $q^i \in \mathrm{Fn}(T^i, 2, \kappa)$ be defined by $q^i(\gamma) = p^i(\alpha_\gamma)$.

Here, we have used the notation $\mathrm{Fn}(T^i, 2, \kappa)$ to denote the set of all partial functions from κ to 2 whose domain is a subset of T^i. So each $\gamma < \kappa^+$, each $\alpha^i_\gamma < \lambda$, each $\theta^i < \kappa^+$, each $T^i \in [i]^{<\kappa}$ and each $q^i \in \mathrm{Fn}(T^i, 2, \kappa) \subsetneq \mathrm{Fn}(i, 2, \kappa)$. For $i \in [\kappa, \kappa^+]$ let $H_i = [i]^{<\kappa} \times \mathrm{Fn}(i, 2, \kappa) \times \kappa$, and write H for H_{κ^+}. Let h^* be an injection from H into κ^+. Define $g : [\kappa, \kappa^+) \longrightarrow \kappa^+$ by setting $g(i)$ to be the least $i^* < \kappa^+$ such that $H_i \subsetneq h^{*-1}``i^*$.

Let $C = \{j < \kappa^+ : (\forall i < j)\theta^i, g(i) < j\}$. As the intersection of the sets of closure points of the two given functions, C is a club subset of κ^+. Let $h(i) = h^*(T^i, q^i, \mathrm{otp}(\mathrm{dom}(p_i))$ for $i \in C \cap S^{\kappa^+}_\kappa$, and $h(i) = 0$ otherwise. We have that $h^{*-1}(h(i)) \in H_i$ for all $i \in [\kappa, \kappa^+)$. If $i \in C \cap S^{\kappa^+}_\kappa$, since $|h^{*-1}(h(i))| < \kappa$, there is some $i' < i$ such that $h^{*-1}(h(i)) \in H_{i'}$, and hence there is some $j < i$ such that $h(i) < g(j)$. Hence, as i is a closure point of g, we have $h(i) < i$ for all non-zero $i < \kappa^+$.

Now suppose that $i, j \in C \cap S^{\kappa^+}_\kappa$, $i < j$ and $h(i) = h(j)$. So we have $T^i = T^j$, $q^i = q^j$ and $\mathrm{otp}(\mathrm{dom}(p_i)) = \mathrm{otp}(\mathrm{dom}(p_j))$. Set $A = \mathrm{dom}(p_i) \cap \mathrm{dom}(p_j)$.

Lemma 16.1.6 *The following hold:*
(a) $(\mathrm{dom}(p_j) \setminus \mathrm{dom}(p_i)) \cap \{\alpha_\gamma : i \leq \gamma < j\} = \emptyset$, *and*
(b) $A \subsetneq \{\alpha_\gamma : \gamma < i\}$.

Proof of 16.1.6 Suppose $\alpha_\gamma \in \mathrm{dom}(p_j)$. If $\gamma < j$ then $\gamma \in T^j$. But $T^j = T^i$, so $\gamma \in T^i$. Hence $\gamma < i$ and $\alpha_\gamma \in \mathrm{dom}(p_i)$, proving (a). If $\alpha_\gamma \in \mathrm{dom}(p_i)$ then $\gamma < \theta^i < j$. (For the definition of θ^i immediately gives that $\gamma < \theta^i$; and since $i < j \in C$ one has that $\theta^i < j$.) Thus if $\alpha_\gamma \in \mathrm{dom}(p_i) \cap \mathrm{dom}(p_j)$ we have $\gamma < i$ by (a). So (b) holds. $\bigstar_{16.1.6}$

By Lemma 16.1.6(b) we have for $\alpha_\gamma \in A$ that $p^i(\alpha_\gamma) = q^i(\gamma) = q^j(\gamma) = p^j(\alpha_\gamma)$, and hence p^i and p^j agree on the intersection of their domains, and thus are compatible conditions.

16.2 A Different View of ccc for a new Forcing Axiom

In [74], Mekler gave a new proof of the consistency of the existence of a universal graph on $\aleph_1$ along with the negation of the continuum hypothesis, a fact which was proved by different means by Shelah in [103]. As part of his argument, Mekler used a characterisation of ccc forcing in terms of every condition being generic over relevant models, mentioned explicitly in [76], which we reproduce now.

Theorem 16.2.1 *A forcing notion $\mathbb{P}$ is ccc iff there is a club set of countable $M \prec \mathcal{H}(\chi)$ such that every $q \in \mathbb{P}$ is $(M, \mathbb{P})$-generic.*

Proof In the forward direction, let $M \prec \mathcal{H}(\chi)$ be countable such that $\mathbb{P} \in M$. Let $q \in \mathbb{P}$ and $r \geq q$ and suppose that $\mathcal{D}$ is an open dense subset of $\mathbb{P}$ with $\mathcal{D} \in M$. Let $\mathcal{A}$ be a maximal antichain within $\mathcal{D}$. By the density of $\mathcal{D}$, the antichain $\mathcal{A}$ is actually maximal for $\mathbb{P}$. By ccc, $\mathcal{A}$ is countable. Hence there is $f : \omega \to \mathcal{D}$ such that ran(f) satisfies that every $p \in \mathbb{P}$ is compatible with an element of ran(f). By elementarity, there is such $f \in M$, which implies that $\mathcal{D} \cap M$ is predense and hence predense above q for any q. Therefore any q is generic.

In the other direction, suppose for a contradiction that there is a maximal antichain $\mathcal{A} = \{p_\alpha : \alpha < \kappa\} \in \mathbb{P}$, where $\kappa \geq \aleph_1$ and that $M \prec \mathcal{H}(\chi)$ is countable, such that $\mathcal{A}, \mathbb{P} \in M$ and that every condition is generic, which can be seen to exist by the assumptions. Let $\delta \in \kappa \setminus M$, which exists since κ is uncountable. In particular, p_δ is supposed to be $(M, \mathbb{P})$-generic.

The set $\mathcal{D} = \{q : (\exists\alpha)\, q \geq p_\alpha\}$ is open dense and in M, hence by genericity, there is $s \in \mathcal{D} \cap M$ which is compatible with p_δ. This implies that there is $\alpha \neq \delta$ such that p_α and p_δ are compatible, which is a contradiction. This shows that $\mathbb{P}$ cannot have a maximal antichain of size $\geq \aleph_1$ and is hence ccc. $\bigstar_{16.2.1}$

Generalising this observation, in the work of Cummings, Džamonja and Neeman [18] we have developed a new framework for strengthening κ^+-cc, which is based on the interaction of the forcing with elementary submodels and which leads to a new forcing axiom. We shall present a simpler version of the axiom from [18], at the price of requiring a stronger closure condition. Namely, calling the above strengthening of κ^+-cc 'strong κ^+-cc' (see Definition 16.3.2), we obtain an iteration theorem for strongly κ^+-cc $(< \kappa)$-strongly directed closed forcing with supports of size $< \kappa$ (Theorem 16.3.12). Here, $(< \kappa)$-*strongly directed closed* forcing means that every directed set of size $< \kappa$ has a least upper bound.

The work in [18] was inspired by the idea of strongly proper forcing, invented by Mitchell in [80] and exploited in the recent work of Neeman, see [85] and [40], in his approach to Baumgartner's problem of the consistency of every two $\aleph_2$-dense sets of reals being isomorphic and the continuum equal at least $\aleph_3$. Our §16.5 is devoted to one of the main results of Neeman's programme, his new proof of the consistency of PFA from a supercompact cardinal, using finite supports and a mixture of properness and strong properness.

16.3 Strong κ^+-cc Forcing without Combinatorics

Let κ be a cardinal satisfying $\kappa = \kappa^{<\kappa}$. We are interested in the case that $\kappa \geq \aleph_1$ and we shall consider elementary submodels M of $\mathcal{H}(\chi) = \langle H(\chi), \in, <^* \rangle$. However, these models will not necessarily be countable, rather they will have cardinality κ. We note that the combinatorics of $[\lambda]^\kappa$ for $\lambda \geq \kappa$, including the club and stationary sets, is analogous to that of $[\lambda]^{\aleph_0}$.

Definition 16.3.1 Let $\mathbb{P}$ be a forcing notion and $M \prec \mathcal{H}(\chi)$.

- We say that M is *adequate for* $\mathbb{P}$, if for every $p, p' \in \mathbb{P} \cap M$

$$p, p' \text{ are compatible in } \mathbb{P} \text{ iff } p, p' \text{ are compatible in } \mathbb{P} \cap M.$$

- Suppose that M is adequate for $\mathbb{P}$. A condition q is $(M, \mathbb{P})$-*strongly generic* (or *a strong master condition*) if for every $r \geq q$, there is a *strong properness residue* $r|M$ of r, which is a condition in M such that

$$(\forall t \in \mathbb{P} \cap M)\, t \geq r|M \implies t, r \text{ are compatible.}$$

We call any such $r|M$ a *strong properness residue* or just a *strong residue* of r in M.

Strongly generic conditions were first used in connection with strong properness, a notion that we discuss in §16.5. Although it came later, the following notion of strongly κ^+-cc from [18] is the easier one to present before the notion of strong properness, since it is in a sense a cc version of strong properness.

Definition 16.3.2 We say that $\mathbb{P}$ is *strongly* κ^+*-cc* if there is a stationary set of M with $|M| = \kappa$, ${}^{<\kappa}M \subseteq M$ and such that $\mathbb{P} \in M$, for which every condition in $\mathbb{P}$ is strongly $(M, \mathbb{P})$-generic. Restricting our attention to a concrete model M we obtain the notion of an M-strongly κ^+-cc forcing. We define the notion of an M-κ^+-cc forcing in an analogous way, by replacing 'strongly $(M, \mathbb{P})$-generic' by '$(M, \mathbb{P})$-generic'.

We note the following:

Lemma 16.3.3 *Suppose that $\mathbb{P}$ is strongly κ^+-cc. Then it is κ^+-cc.*

Proof Let $\mathbb{P}$ be strongly κ^+-cc. Suppose that $\bar{p} = \langle p_i : i < \kappa^+ \rangle$ is a sequence of conditions in $\mathbb{P}$ and suppose for a contradiction that this sequence is an antichain. Let us choose $M \prec \mathcal{H}(\chi)$ such that $\kappa, \mathbb{P}, \bar{p} \in M$ and ${}^{<\kappa}M \subseteq M$ while $|M| = \kappa$ and such that $\mathbb{P}$ is strongly M-κ^+-cc, which exists by an application of stationarity. It follows that $\kappa^+ \cap M = \delta$ is some ordinal. Let $r = p_\delta|M$. In particular r is compatible with some p_j for a $j < \kappa^+$, namely p_δ. By elementarity, there is $i < \delta$ such that p_i and r are compatible. Then there is $s \in M$ such that $s \geq p_i, r$. By the definition of $p_\delta | M$, conditions s and p_δ are compatible, in particular p_i and p_δ are compatible, which is a contradiction. $\bigstar_{16.3.3}$

We show that Cohen forcing is strongly κ^+-cc. Work in progress is to find more examples of strongly κ^+-cc forcing. One interesting candidate which seems to be strongly κ^+-cc is Mekler's forcing from [74].

Theorem 16.3.4 *The forcing* Add(κ, λ) *to add* λ *Cohen subsets to* κ *by conditions of size* $< \kappa$ *is strongly* κ^+*-cc.*

Proof Let $M \prec H(\chi)$, $|M| = \kappa$, ${}^{<\kappa}M \subseteq M$ and $\kappa, \lambda \in M$. Then for any condition r in Add$(\kappa, \lambda,)$, it suffices to let $r|M = r \restriction (\text{dom}(r) \cap M)$. $\bigstar_{16.3.4}$

In fact, as Mitchell explains in [80], the idea of strongly generic conditions is that above such a condition, the forcing splits into a two step iteration, see [80], which of course fits with the properties of Cohen's forcing.

We prove that under reasonable closure properties, strongly κ^+-cc forcing can be iterated with $(< \kappa)$-supports. To recall, this means that we shall henceforth be concerned with forcing notions of the form $\mathbb{P}$ where $\mathbb{P}$ is the result of an iteration

$$\langle P_\alpha, \underset{\sim}{Q}_\beta : \alpha \leq \alpha^*, \beta < \alpha^* \rangle$$

where every condition p satisfies $|\text{supt}(p)| < \kappa$.

We start with the notion of the least upper bound (denoted lub), whose definability makes it convenient in iterations, as noted in Observation 16.3.5.

Observation 16.3.5 *Suppose that* $\mathbb{P}$ *is the result of an iteration of length* α^*, $A \subseteq \mathbb{P}$ *and* $q = \text{lub}(A)$. *Then for every* $\alpha < \alpha^*$, *it is forced by* $q \restriction \alpha$ *that* $g(\alpha)$ *is the lub of* $\{p \restriction \alpha : p \in A\}$.

The closure property used in Theorem 16.3.12 is given by the following definition.

Definition 16.3.6 We say that $\mathbb{P}$ is $(< \kappa)$*-strongly directed closed* if every directed sequence of length $< \kappa$ and consisting of conditions in $\mathbb{P}$, has a least upper bound.

It is well known (see [99] or [17] for an explanation, for example) that the notion of $(< \kappa)$-strong directed closure is preserved under the iteration with $(< \kappa)$-supports of κ^+-cc forcing that preserve cardinals. We shall sometimes use the following consequence of the $(< \kappa)$-strong directed closure.

Observation 16.3.7 *Suppose that* $\mathbb{P}$ *is a* $(< \kappa)$*-strongly directed closed forcing,* $\langle p_j : j < \omega \rangle$ *is an increasing sequence and* s *a condition which is compatible with each* p_j. *Then the set* $\{p_j : j < \omega\} \cup \{s\}$ *has the least upper bound.*

Proof It suffices to note that the set $\{p_j : j < \omega\} \cup \{s\}$ is directed, which follows from the assumptions, and that it clearly has size $< \kappa$. $\bigstar_{16.3.7}$

16.3.1 Canonical Extensions

We first observe an evident, but useful property of strong residues.

Observation 16.3.8 *Suppose that $r \geq q$ in some forcing $\mathbb{P}$ and that M is a model such that there is a strong residue $r|M$. Then $r|M$ is a strong residue for q in M.*

In particular, to show that a forcing notion is strongly κ^+-cc, it suffices to show that a dense set of conditions in it admits strong residues in appropriate models.

The following process will be called 'constructing a *canonical extension*'.

Definition 16.3.9 Suppose that $\mathbb{P}$ and M are appropriate. A *canonical extension* $q \geq p$, if it exists, is defined by constructing sequences $\langle p_i : i < \omega \rangle$ and $\langle H_i : i < \omega \rangle$ so that

1. for all i, $H_i \prec H(\chi)$ and $|H_i| < \kappa$,
2. $p, M \in H_0$ and $\mathrm{supt}(p) \subseteq H_0$,
3. $p_0 = p$,
4. for all i, $p_{i+1} \geq p_i$ and $p_{i+1} \in \mathcal{D}$ for every open dense $\mathcal{D} \in H_i$,
5. for all i, $H_i \cup \{p_{i+1}, H_i\} \cup \mathrm{supt}(p_{i+1}) \subseteq H_{i+1}$,
6. $\langle p_i : i < \omega \rangle$ admits a least upper bound,

and then letting $q = \mathrm{lub}_{i<\omega}\, p_i$. Let $H = \bigcup_{i<\omega} H_i$.

Lemma 16.3.10 *Suppose that q is a canonical extension of p as in Definition 16.3.9. Then the following are true:*

1. *$H \prec H(\chi)$, $|H| < \kappa$ and $p, M \in H$,*
2. *if $g = \{s \in \mathbb{P} \cap H : (\exists i < \omega)\, s \leq p_i\}$ then q is the lub of g and $g = \{s \in \mathbb{P} \cap H : s \leq q\}$,*
3. *g is a filter on $\mathbb{P} \cap H$ which meets every open set in H that is dense above some p_i,*
4. *$\mathrm{supt}(q) = H \cap \gamma = \bigcup_{i<\omega} \mathrm{supt}(p_i)$,*
5. *$H \cap M \in M$.*

Proof The first item is obvious. For the second one, clearly every upper bound r of g is an upper bound of all p_i, so $r \leq q$. On the other hand, q is an upper bound for g, so it follows that q is a lub for g. It is similarly proved that $g = \{s \in \mathbb{P} \cap H : s \leq q\}$.

Since the sequence $\langle p_i : i < \omega\rangle$ is increasing we have that g is a filter. Every open dense subset of $\mathbb{P}$ that is in H is in some H_i, hence it contains some p_{i+1} and therefore it meets g. If an open set $\mathcal{D} \in H$ is dense above some p_i, then the set $\mathcal{D}' = \{t : t \perp p_i \vee t \in \mathcal{D}\}$ is open dense and in H. Hence $\mathcal{D}'$ meets g. By the definition of g, all elements of $g \cap \mathcal{D}'$ must actually come from $\mathcal{D}$, therefore g meets $\mathcal{D}$.

By the two paragraphs above it follows that $\mathrm{supt}(q) \supseteq H \cap \alpha^*$ and the fact that $\mathrm{supt}(q) \subseteq H \cap \alpha^*$ follows from the fact that q is the $\mathrm{lub}(\langle p_i : i < \omega\rangle)$ and Observation 16.3.5. The last item follows because $|H \cap M| < \kappa$ and the fact that M is closed under $(< \kappa)$-sequences. ★16.3.10

By considering the definition of canonical extension, we note:

Observation 16.3.11 *Any $(< \kappa)$-strong directed closed forcing admits a construction of a canonical extension $q \geq p$ for any $p \in \mathbb{P}$.*

16.3.2 The Iteration Theorem

Theorem 16.3.12 *Suppose that $M \prec H(\chi), |M| = \kappa$ and ${}^{<\kappa}M \subseteq M$. Every $\mathbb{P} \in M$ obtained as an iteration with supports of size $(< \kappa)$ of strongly M-κ^+-cc $(< \kappa)$-strongly directed closed forcing, is itself strongly M-κ^+-cc.*

Proof We need to show that a dense set of conditions in $\mathbb{P}$ has a strong residue over M. Let $p \in \mathbb{P}$ be arbitrary. The assumptions on the iterands imply that $\mathbb{P}$ is $(< \kappa)$-strongly directed closed, so by Observation 16.3.11, we can construct a canonical extension $q \geq p$. We use the same notation as in Definition 16.3.9 and in particular have the set g as defined there.

Consider now the set $g \cap M$, which is a directed subset of $\mathbb{P}$ and has size $< \kappa$, since $g \subseteq H$ and $|H| < \kappa$. Let $t = \mathrm{lub}(g \cap M)$, which exists since $\mathbb{P}$ is strongly $(< \kappa)$-directed closed. Since M is closed under $(< \kappa)$-sequences, we have that $g \cap M \in M$, so by the definability of lub we conclude that $t \in M$. We claim that t is a strong residue for q in M.

First, since q is an upper bound for g, in particular q is an upper bound for $g \cap M$ and hence $q \geq t$, since t is the least upper bound. Suppose now that $s \in M$ and $s \geq t$, we need to show that s is compatible with q. We shall build a common extension q^* of s and q by inductively defining $q^* \restriction \alpha$ for $\alpha \leq \alpha^*$. Having defined $q^* \restriction \alpha$, we show how to define $q^*(\alpha)$, as a name over $q^* \restriction \alpha$. For notational convenience, assume that $\alpha^* = \alpha + 1$ here, as the tail of the iteration does not influence this construction at α.

If α is not in the support of both q and s, we simply choose $q^*(\alpha)$ to be either $s(\alpha)$, $q(\alpha)$ or the trivial condition, as appropriate. Now suppose that α is in the support of both q and s. We may as well assume that $\alpha \in \text{supt}(p_i)$ for all i, since otherwise we can throw away a few first elements of $\langle p_i : i < \omega \rangle$. Since $s \in M$ we have $\text{supt}(s) \subseteq M$ and hence $\alpha \in M$. On the other hand, since q is a canonical extension of p obtained through building H, we have by Lemma 16.3.10(4) that $\text{supt}(q) \subseteq H$ and so $\alpha \in H \cap M$. We now define certain subsets of $\mathbb{P}$, using only parameters in H, hence each of them will be an element of H. For each $i < \omega$, let $\mathcal{D}_i$ be the set of all $u \in \mathbb{P}$ such that, using s.r. for a 'strong residue':

$$(\exists \underset{\sim}{r}^* \in M \cap \underset{\sim}{Q}_\alpha)\ u \restriction \alpha \Vdash_\alpha \text{`}u(\alpha) \geq \underset{\sim}{r}^* \text{and } \underset{\sim}{r}^* \text{ is s. r. for } p_i(\alpha) \text{ in } M[\underset{\sim}{G}_\alpha]\text{'}.$$

Clearly, for each i the set $\mathcal{D}_i$ is open. We claim that $\mathcal{D}_i$ is dense above p_i. To see this, fix $i < \omega$ and some condition $v \geq p_i$. Since $\underset{\sim}{Q}_\alpha$ is forced by P_α to be strongly κ^+-cc and since by the inductive hypothesis P_α is strongly κ^+-cc, we have that $M[\underset{\sim}{G}_\alpha]$ is forced to be an elementary submodel of $H_\chi([\underset{\sim}{G}_\alpha])$ of size $< \kappa$ and closed under $(< \kappa)$-sequences. Equally, $v \restriction \alpha$ has an extension in P_α which decides a strong residue for $v(\alpha)$ in $M[\underset{\sim}{G}_\alpha]$. Without loss of generality, v is already such a condition and let $\underset{\sim}{z}$ be such that $v \restriction \alpha$ forces $\underset{\sim}{z}$ to be a strong residue for $v(\alpha)$ in $M[\underset{\sim}{G}_\alpha]$. In particular $v \restriction \alpha$ forces $\underset{\sim}{z}$ to be compatible with $v(\alpha)$ so we can find $v' \geq v$ such that $v' \restriction \alpha = v \restriction \alpha$ and $v' \restriction \alpha \Vdash_\alpha$ '$v'(\alpha) \geq \underset{\sim}{z}$'. Since $v' \geq v \geq p_i$ and $v' \restriction \alpha \Vdash_\alpha$ '$\underset{\sim}{z}$ is a strong residue for $v(\alpha)$', we have that $v' \in \mathcal{D}_i$.

We may conclude that g meets each $\mathcal{D}_i$. By induction on $k < \omega$ we shall construct sequences $\langle i_k : k < \omega \rangle$, $\langle q_k : k < \omega \rangle$ and $\langle \underset{\sim}{r}^*{}_k : k < \omega \rangle$ such that

- $i_0 = 0$,
- $q_k \in \mathcal{D}_{i_k} \cap g$ as exemplified by $\underset{\sim}{r}^*{}_k$ and $\underset{\sim}{r}^*{}_k \in H$,
- i_{k+1} is such that $p_{i_{k+1}} \geq q_k$.

The construction of these sequences uses the definition of g and, to be able to choose $\underset{\sim}{r}^*{}_k \in H$, the elementarity of H and the fact that α, M and $p_{i_k} \in H$. In particular, $\underset{\sim}{r}^*{}_k \in M \cap H$, for all k. Let r_k be the condition with $r_k \restriction \alpha$ trivial and with $r_k(\alpha) = \underset{\sim}{r}^*{}_k$. We have that $r_k \in M \cap H$ and that $q_k \geq r_k$, so in particular $q \geq r_k$ for all k and $q \restriction \alpha \Vdash_\alpha$ '$\underset{\sim}{r}^*_k$ is a strong residue for $p_{i_k}(\alpha)$ in $M[\underset{\sim}{G}_\alpha]$'. Since $q \geq p_{i_{k+1}} \geq q_k$, and p_is are increasing, we have that $q \restriction \alpha \Vdash_\alpha$ '$p_i(\alpha), s(\alpha)$ are compatible', for all i. Using Observation 16.3.7 on $\underset{\sim}{Q}_\alpha$ over P_α and on the sequence $\langle p_i(\alpha) : i < \omega \rangle$ and $s(\alpha)$, we obtain that $q^* \restriction \alpha$ forces that there is an upper bound for all $p_i(\alpha)$ and $s(\alpha)$ in $\underset{\sim}{Q}_\alpha$, which we then take as $q^*(\alpha)$. It is clear that $q^* \restriction \alpha \Vdash_\alpha$ '$q^*(\alpha) \geq s(\alpha)$', but it is also true that $q^* \restriction \alpha \Vdash_\alpha$ '$q^*(\alpha) \geq q(\alpha)$' because q is the lub of g and $q^* \restriction \alpha$ forces that $q^*(\alpha)$ is an upper bound for $\langle p_i(\alpha) : i < \omega \rangle$.

★16.3.12

16.4 Method of Side Conditions

Todorčević's paper [117] where he proved that PFA implies that $\square_\kappa$ fails for all uncountable κ, introduced a new way of looking at proper forcing, namely as forcing with *side conditions*. This method has since been used to prove many theorems and forms a basis for a later development by Neeman [85] in which a new proof of consistency of PFA is obtained. In this section we explain Todorčević's proof and in the next we give Neeman's result.

Theorem 16.4.1 (Todorčević) *Suppose that PFA holds and let $\kappa > \aleph_1$ be regular. Then, for any sequence $\langle C_\delta : \delta \text{ limit } < \kappa\rangle$ such that*

- *C_α is a club of α,*
- *if β is a limit point of C_α, then $C_\beta = C_\alpha \cap \beta$,*

there is a club C of κ such that for all α limit, we have $C_\alpha = C \cap \alpha$.

As a consequence, under PFA, the principle $\square_\lambda$ fails for any λ uncountable.

Proof First let us prove the consequence. Suppose that the theorem is true yet a sequence $\langle C_\alpha : \alpha \text{ limit } < \lambda^+\rangle$ is a $\square_\lambda$-sequence for some λ uncountable. Let $\kappa = \lambda^+$ and we apply the conclusion of the theorem to find a club C of κ such that for all α limit we have $C_\alpha = C \cap \alpha$. On the other hand, the definition of $\square_\lambda$ implies that $\mathrm{otp}(C_\alpha) < \alpha$ for all $\alpha > \omega$, so the regressive function $\mathrm{otp}(C_\alpha)$ must be constant on a stationary set S. Since $S \cap C$ is stationary, we can find two elements $\alpha < \beta$ of $S \cap C$. Therefore we have $C \cap \alpha = C_\alpha$ and $C \cap \beta = C_\beta$. Since $\mathrm{otp}(C_\alpha) = \mathrm{otp}(C_\beta)$, it must be that $C \cap [\alpha, \beta) = \emptyset$, which is a contradiction since $\sup(C_\beta) = \beta$.

Let us now prove the theorem. Suppose that $\kappa > \aleph_1$ and $\bar{C} = \langle C_\delta : \delta \text{ limit } < \kappa\rangle$ is given. The first step is to consider the C_δ's as forming a tree of the tries at an initial segment of the future C. So we let $\alpha <_T \beta$ if C_α is an initial segment of C_β, which by the assumption happens iff α is a limit point of C_β. This order forms a tree T and any κ-branch of that tree will give a C as required. Now comes the main idea.

Todorčević's method of models as side conditions We shall consider elementary chains of length ω_1 of countable models $N \prec \mathcal{H}(\kappa^+)$ (note: we consider κ^+ here and not χ; the reader will see why when we get to the proof of properness in Claim 16.4.2). To be precise, we shall consider sequences of the form $\bar{N} = \langle N_\alpha : \alpha \in A\rangle$ for some subset A of ω_1, such that

1. $\bar{C} \in N_\alpha$, $N_\alpha \prec \mathcal{H}(\kappa^+)$ and $|N_\alpha| = \aleph_0$,
2. if $\alpha < \beta$ are both in A, then $N_\alpha \in N_\beta$,
3. for any δ which is a limit point of A, we have $N_\delta = \bigcup_{\alpha<\delta} N_\alpha$.

For any $N \prec \mathcal{H}(\kappa^+)$ countable we let $\kappa_N = \sup(N \cap \kappa)$. Note that each κ_N is a limit ordinal and therefore C_{κ_N} is defined. We consider *partial specialising functions* $f : F \to \omega$, where $F \subseteq \kappa$ is a finite set of limit ordinals and for $\alpha, \beta \in \text{dom}(F)$,

$$\alpha <_T \beta \implies f(\alpha) \neq f(\beta).$$

We now define a forcing notion $\mathbb{P}$. Each condition will have a finite *working part*, which is a partial specialising function and a *countable models part* which will be a finite elementary chain of countable models. Precisely, $\mathbb{P}$ consists of conditions of the form $(f_p, \bar{N}_p)$ where:

- $\bar{N}_p = \langle N^p_\alpha : \alpha \in A_p \rangle$ is an elementary chain indexed by a finite $A_p \subseteq \omega_1$ but such that it can be extended to an elementary chain indexed by ω_1,
- f_p is a finite specialising function such that $\text{dom}(f) \subseteq \{\kappa_{N^p_\alpha} : \alpha \in A_p\}$.

We order $\mathbb{P}$ by letting $p \leq q$ iff $A_p \subseteq A_q$, $N^q_\alpha = N^p_\alpha$ for $\alpha \in A^p$ and $f_p \subseteq f_q$.

Claim 16.4.2 $\mathbb{P}$ is proper.

Proof of Claim 16.4.2 Let $p \in \mathbb{P}$ and $M \prec \mathcal{H}(\chi)$ countable such that $p, \mathbb{P} \in M$. Let $\delta = M \cap \omega_1$. As often in the constructions of generic conditions, to extend p to a generic q we will consider some action on δ. In fact, we shall simply add a model at the empty δ spot of the sequence $\bar{N}^p$, and that model will be $M \cap \mathcal{H}(\kappa^+)$. The reader may verify that $\bar{N}^p \frown (\delta, M \cap \mathcal{H}(\kappa^+))$ satisfies the requirements of being a model part of a condition in $\mathbb{P}$ which is above p. For example, if $\alpha \in A_p$, we know that $\alpha \in M$ and since $\bar{N}^p \in M$ we have $N^\alpha_p \in M \cap \mathcal{H}(\kappa^+)$. So let

$$q = (f^p, \bar{N}^p \frown (\delta, M \cap \mathcal{H}(\kappa^+))).$$

Clearly $p \leq q$ and we need to check that q is $(M, \mathbb{P})$-generic. Let $r \geq q$ and let $\mathcal{D} \in M$ be an open dense subset of $\mathbb{P}$. We may assume $r \in \mathcal{D}$. Usually in the proofs of genericity, at this point we take some sort of a 'shadow' or a 'residue' r' of r within M. In this case we can literally define $r' = r \cap M$. It is important to check that this is really a condition and that it is in M. In fact, we need to note that $r \cap M = r \cap M \cap \mathcal{H}(\kappa^+)$, since the whole forcing is a subset of $\mathcal{H}(\kappa^+)$. Therefore the models in $\bar{N}^{r'}$ are exactly those models from $\bar{N}^r$ which are in $M \cap \mathcal{H}(\kappa^+)$ (using the fact that $r \geq q$ and that $M \cap \mathcal{H}(\kappa^+) \in \bar{N}^q$), and $f^{r'}$ is $f^r \restriction M$, while for every $N \in M$ we have $\kappa_N \in M$. Therefore r' is a well defined condition and it is in M. It is also easy to check that $r' \geq p$ while $r' \leq r$.

Now we shall try to extend r' into $M \cap \mathcal{D}$ while keeping it compatible with r. Rather than constructing such an extension directly, we shall make many guesses and then show that one of them must work. To make these guesses, we shall 'imitate' r, as much as we can. Let $F = \mathrm{dom}(fr) \setminus M$ and let $n = |F|$. Clearly $n \geq 1$ since $\kappa_M \in \mathrm{dom}(f^r)$. Within M we can construct sequences $\langle F_\xi : \xi < \kappa\rangle^M$ and $\langle r_\xi : \xi < \kappa\rangle^M$ so that:

- $F_\xi \subseteq \kappa, |F_\xi| = n$,
- $\xi < \zeta \implies (\forall \alpha \in F_\xi)(\forall \beta \in F_\zeta)\, \alpha < \beta$,
- $r_\xi \in \mathcal{D}, r_\xi \geq r'$ and $\mathrm{dom}(f^{r_\xi}) = \mathrm{dom}(f^{r'}) \cup F_\xi$.

To see this, if we have already constructed $\langle F_\xi : \xi < \zeta\rangle^M$ and $\langle r_\xi : \xi < \zeta\rangle^M$, we let $\zeta^* = \sup^M(\bigcup_{\xi<\zeta} F_\xi)$, which is $< \kappa_M$ by the fact that $M \models \mathrm{cf}(\kappa) = \kappa$. Then we consider the following sentence φ with the variables r'', F and parameters in M:

$$(\exists F \in [\kappa]^n)(\exists r'' \in \mathbb{P})\ \min(F) > \zeta^* \wedge \mathrm{dom}(f^{r''}) = \mathrm{dom}(f^{r'}) \cup F \wedge r'' \in \mathcal{D} \wedge r'' \geq r'.$$

Then $\mathcal{H}(\chi)$ satisfies the sentence, as exemplified by r and F, and therefore there are r'', F in M that satisfy φ, so we can choose them as r_ζ and F_ζ.

Within M we have that T is a tree of size and height κ which has no κ-branch. By methods similar to the proof of Lemma 12.1.6, it follows that there must be $\xi \in M$ such that no element of F_ξ is comparable to any element of F. This means the function $f^{r_\xi} \cup f^r$ is still a specialising function and then it is easy to check that $s = (f^{r_\xi} \cup f^r, \bar{N}^{r_\xi} \cup \bar{N}^r)$ is a condition which is a common extension of r and r_ξ. Knowing that $r_\xi \in \mathcal{D} \cap M$, this finishes the claim. ★16.4.2

We now note that $\mathbb{P}$ has a dense set of conditions p with

$$\mathrm{dom}(f^p) = \{\kappa_N : N \in \bar{N}^p\}$$

and we shall apply PFA to the induced poset, which we shall for simplicity again call $\mathbb{P}$. For every $\alpha < \omega_1$, the set $\mathcal{D}_\alpha$ of conditions p such that $\alpha \in A_p$ is dense (by a Löwenheim–Skolem argument). Let G be a filter that intersects all $\mathcal{D}_\alpha$, then $\bar{N} = \bigcup_{p\in G} \bar{N}^p$ is an elementary ω_1-chain of models, which we shall denote by $\langle N_\alpha : \alpha < \omega_1\rangle$. We shall define $\kappa_\alpha = \sup(N_\alpha \cap \kappa)$ and let $\gamma = \sup_{\alpha<\omega_1} \kappa_\alpha$.

On the other hand, let $f = \bigcup_{p\in G} f^p$, which gives a specialising function $f : \{\kappa_\alpha : \alpha < \omega_1\} \to \omega$. Now note that $\{\kappa_\alpha : \alpha < \omega_1\}$ and the set $\lim(C_\gamma)$ of limit points of C_γ are both clubs of γ and hence their intersection D is as well. By the definition of $\square_\kappa$ we have that for all $\kappa_\alpha \in D$, $C_{\kappa_\alpha} = C_\gamma \cap \kappa_\alpha$ and in particular $\langle \kappa_\alpha : \kappa_\alpha \in D\rangle$ forms a chain in $<_T$. However this is impossible since f is 1–1 on any chain and the range of f is ω. ★16.4.1

An analysis of the above proof shows that rather than working with sequences of clubs defined for all limit ordinals, it was sufficient to consider any set of limit ordinals which contains $S^{\kappa^+}_{\omega_1}$, which is how the original theorem is phrased in [117].

16.5 Side Conditions for ω_2

The method of side conditions found many applications in proving combinatorial results on ω_1. The one that turned out the most relevant to the idea of generalising proper forcing to higher cardinals was obtained by Baumgartner in [§3 [6]], where a club is added to ω_1 using finite approximations, but side conditions are used to control the closure at the limit points. Pioneering work on applying the method of side conditions to the combinatorics of ω_2 was done by Piotr Koszmider [59] who proved that it is consistent to have a sequence of length ω_2 increasing modulo finite in ${}^{\omega_1}\omega_1$, but the method received more attention with the work of Sy D. Friedman [36] and Mitchell [79], who independently found a way to add a club to ω_2 using finite conditions and side models. In fact such a thing was already done in the 1990s by Andrzej Rosłanowski and Shelah in [93], but their work went largely unnoticed, perhaps due to its rather technical nature. Maybe it would have been the case with the work in [36] and [79] too, had it not been that Mitchell's work solved an important open problem (about the consistency of the triviality of a certain ideal $I[\aleph_2]$ on ω_2), by iterating adding such clubs, although his method was very difficult to parse. Gregor K. Dolinar and Džamonja [23] went one step further in adding a single object to ω_2, in that they were able to add a $\square_{\omega_1}$-sequence by using finitely many models on each condition but using countable initial segments of the clubs. That development followed Mitchell's method and again was very technical. A posteriori, looking closely at the second example in §3 of [93], one can see that some sort of square is added by small conditions, but this is not stated explicitly. All of the above methods worked either without side conditions such as [93], or with a sequence of elementary submodels as a side condition, plus some extra combinatorial requirements such as the idea of fences in [79] and [23].

Neeman [85] made a breakthrough development in understanding that what was needed is not one sequence of models, but two sequences of *different* models. In this way he was able to give a new proof of the consistency of PFA (again from a supercompact cardinal) and to get considerably simpler proofs of Koszmider's result from [59] and the Dolinar–Džamonja result from [23]. Neeman's work inspired other researchers to develop this method, for example John Krueger (such as [60] on adding a square). Neeman and others obtained many other applications of the method, see Neeman's paper with

Thomas Gilton [40] and Veličković's paper with Giorgio Venturi [122] for further examples. Neeman's methods give a hope of having a workable version of PFA for $\aleph_2$–although it would have to be a much weaker axiom than its analogue on $\aleph_1$, as we shall explain in §16.6 –and a more palpable goal of obtaining the consistency of every two $\aleph_2$-sets of reals being isomorphic and the continuum $\aleph_3$, where Neeman has a solution in preparation. Let us explain his method of working with two kinds of models as side conditions. Although the method applies in a more general situation, we shall present the simplest case which is already sufficient for our purposes.

Two kinds of models side conditions. The first kind of models are *small models*, which are countable $M \prec \mathcal{H}(\chi)$, forming a collection $\mathcal{S}$. The second kind, *transitive models* are of the form $W \prec \mathcal{H}(\chi)$ with $W \in \mathcal{H}(\chi)$ transitive, forming a collection $\mathcal{T}$. The connection required from $\mathcal{S}$ and $\mathcal{T}$ is that

$$(\forall M \in \mathcal{S})(\forall W \in \mathcal{T})\, W \in M \implies M \cap W \in \mathcal{S} \text{ and } M \cap W \in W.$$

This condition will be satisfied if, for example, all models in $\mathcal{T}$ are countably closed and $\mathcal{S}$ collects all countable elementary submodels of $\mathcal{H}(\chi)$. A *side condition* then is a finite sequence $\langle M_0 \in M_1 \in \ldots \in M_n \rangle$ of models from $\mathcal{S} \cup \mathcal{T}$, which is closed under intersections. These side conditions form a forcing notion of their own, when ordered by $\subseteq$.

Definition 16.5.1 If s is a side condition and Q one of the models appearing in s, we call such a model a *node*, and we distinguish small nodes and transitive nodes. For any node, we define the *residue* of s in Q, denoted by $r_s(Q)$ as the subsequence of s consisting of all models in s that contain Q.

The following are the most important properties of the side conditions poset. See §2 and §3 of [85] for the proofs.

Lemma 16.5.2 *If s is a side condition and Q one of the models appearing in s, then:*

- *$r_s(Q)$ is a side condition and*
- *if $t \in Q$ is a side condition and $r_s(Q) \leq t$, then s and t are compatible.*

The reader may realise that the second item of Lemma 16.5.2 looks like the arguments in proper forcing, when we are looking for a residue within a model of a condition that is above the generic condition. This is the right intuition, because not only is the side condition poset proper, but it is *strongly proper*, in the sense of the following definition, due to Mitchell in [79].

Definition 16.5.3 Let $\mathbb{P}$ be a forcing notion and $M \prec \mathcal{H}(\chi)$. We say that $\mathbb{P}$

is *strongly proper* if for every M that is adequate for $\mathbb{P}$, for every $p \in \mathbb{P} \cap M$, there is $q \geq p$ which is $(M, \mathbb{P})$-strongly generic.

Note that the notion of M being adequate for $\mathbb{P}$ is weaker than $P \in M$, in particular the notion of strongly proper forcing requires the existence of strongly proper conditions for a larger class of models than the one we are used to dealing with in the arguments about properness.

The above notions allowed Neeman to give a new proof of the consistency of PFA from a supercompact cardinal, without using the preservation of properness by countable support iterations. In fact, his iteration is a finite support iteration, which is completely counter-intuitive by what was known about proper forcing (see Example 15.4.4). It is the side conditions that assure properness. We shall sketch the main point of Neeman's proof from [85].

Let κ be a supercompact cardinal and let f: $\kappa \to \mathcal{H}(\kappa)$ be a Laver diamond (see Definition 13.2.4). We shall define an iteration which will preserve $\aleph_1$ and collapse κ to $\aleph_2$. For our choice of the classes $\mathcal{S}$ and $\mathcal{T}$ as in the description of two kinds of side models, we shall work with $\chi = \kappa$. We shall let $\mathcal{S}$ be the set of all countable $M \prec (H(\kappa); f)$ (his notation means that we are taking f as a distinguished unary function in our model). To define $\mathcal{T}$ we first let

$$Z = \{\alpha < \kappa : (\mathcal{H}(\alpha); f \restriction \alpha) \prec (\mathcal{H}(\chi); f)\}.$$

It is standard to check, using the large cardinal strength of κ, that Z is a club of κ. Let $\mathcal{T}$ be the set of all $W = \mathcal{H}(\alpha)$ for $\alpha \in Z$ of uncountable cofinality. Then each $W \in \mathcal{T}$ is countably closed. Families $\mathcal{S}$ and $\mathcal{T}$ are stationary in $\mathcal{P}_\kappa\kappa$ and they satisfy the requirements for two kinds of models side conditions. Now we describe the forcing, which to honour Neeman's notation, we denote by $\mathbb{A}$.

For each $\alpha \in Z$, define $\lambda(\alpha)$ to be the least cardinal $< \kappa$ such that $f(\alpha) \in \mathcal{H}(\lambda(\alpha))$. It follows from the definition of Z that $\lambda(\alpha) < \min(Z \setminus \alpha)$.

Definition 16.5.4 (1) Conditions in $\mathbb{A}$ are pairs (p, s) such that:

1. s is a two kinds of models side condition,
2. p is a partial function whose domain is contained in the finite set

$$\{\alpha < \kappa : \mathcal{H}(\alpha) \in s \text{ and } \Vdash_{\mathbb{A} \cap \mathcal{H}(\alpha)} \text{`}f(\alpha) \text{ is a proper forcing notion'}\},$$

3. for $\alpha \in \mathrm{dom}(p)$ we have $p(\alpha) \in \mathcal{H}(\lambda(\alpha))$,
4. for $\alpha \in \mathrm{dom}(p)$ we have $\Vdash_{\mathbb{A} \cap \mathcal{H}(\alpha)} \text{`}p(\alpha) \in f(\alpha)\text{'}$,
5. if $M \in s \cap \mathcal{S}$ and $\alpha \in \mathrm{dom}(p) \cap M$,

$$(p \restriction \alpha, s \cap \mathcal{H}(\alpha)) \Vdash_{\mathbb{A} \cap \mathcal{H}(\alpha)} \text{`}p(\alpha) \text{ is } (M, f(\alpha)) - \text{generic'}.$$

(2) The ordering on $\mathbb{A}$ is $(p, s) \leq (q, t)$ if $s \subseteq t$, $\mathrm{dom}(p) \subseteq \mathrm{dom}(q)$ and for each $\alpha \in \mathrm{dom}(p)$, we have

$$(q \restriction \alpha, t \cap H(\alpha)) \Vdash_{\mathbb{A} \cap \mathcal{H}(\alpha)} \text{`}p(\alpha) \leq_{f(\alpha)} q(\alpha)\text{'}.$$

(3) For $\alpha \in Z \cup \{\kappa\}$ let $\mathbb{A}_\alpha$ consist of all conditions (p, s) such that $\mathrm{dom}(p) \subseteq \alpha$, with the order inherited from $\mathbb{A}$.

To make sense of the above definition (1)–(2), one has to note that by the choice of Z, the intersection $\mathbb{A} \cap \mathcal{H}(\alpha)$ is indeed a forcing notion. So although the iteration is not written in the same way as a classical notion of iterated forcing, such as in Theorem 15.4.1, this is still a definition by induction on α and the final step corresponds to $\alpha = \kappa$. Now we shall quote some of the main lemmas from Neeman's proof. The first one shows the interplay between the strong properness and the transitive nodes.

Lemma 16.5.5 *Let $\alpha < \beta$, both in $Z \cup \{\kappa\}$. Suppose that $(p, s) \in \mathbb{A}_\beta$. Then:*

- *if $\mathcal{H}(\alpha) \in s \cap \mathcal{T}$, then s has a strong properness residue in $\mathcal{H}(\alpha)$ in the side conditions forcing, which is $s \cap \mathcal{H}(\alpha)$,*
- *if $\mathcal{H}(\alpha) \in s \cap \mathcal{T}$, then (p, s) is compatible in $\mathbb{A}_\beta$ with any $(q, t) \in A \cap \mathcal{H}(\alpha)$ which extends $(p \restriction \alpha, s \cap \mathcal{H}(\alpha))$ and this is witnessed by $(q \frown p \restriction [\alpha, \beta), s \cup t)$,*
- *if $\mathcal{H}(\gamma)$ is a transitive node in s, then (p, s) is strongly $(\mathcal{H}(\gamma), \mathbb{A}_\beta)$-generic (note that possibly $\gamma \geq \beta$),*
- *if $\gamma \in Z$ is of uncountable cofinality and $(p, s) \in \mathcal{H}(\gamma)$, then $(p, s \cup \{\mathcal{H}(\gamma)\})$ is a condition in $\mathbb{A}_\beta$ and it extends (p, s),*
- *$\mathbb{A}_\beta$ is strongly proper for models in $\mathcal{T}$.*

The second lemma deals with countable models, over which we do not get strong properness, but we do get properness. There is a similar trick as in Todorčević's proof, in that we consider countable submodels of $\mathcal{H}(\theta)$ for some $\theta > \kappa$ (in Todorčević's proof, the analogous point is the use of $\chi > \kappa^+$).

Lemma 16.5.6 *Let $\beta \in Z \cup \{\kappa\}$ and suppose that $(p, s) \in \mathbb{A}_\beta$ and $M^* \prec \mathcal{H}(\theta)$ for some $\theta > \kappa$ is countable such that $\kappa, f, \beta \in M^*$. Let $M = M^* \cap \mathcal{H}(\kappa)$ and suppose that $M \in s$. Then (p, s) is $(M^*, \mathbb{A}_\beta)$-generic.*

These two lemmas are the basis for the proof of the main theorem, namely

Theorem 16.5.7 (Neeman) *Forcing with $\mathbb{A}$ over a model in which κ is a supercompact cardinal preserves $\aleph_1$ and collapses κ to become $\aleph_2$. In addition, PFA holds in the resulting universe.*

16.6 Some Challenges

The same proof that shows the consistency of $MA(\omega_1)$ and the negation of CH can be used to show for example the consistency of $MA(\omega_2)$ and $2^{\aleph_0} = \aleph_3$ and so on. This is however very much in contrast with the case of PFA, which as we mentioned above implies that $2^{\aleph_0} = \aleph_2$ and therefore it is hopeless to obtain PFA for families of $\aleph_2$ dense sets. Neeman's new method suggests that it should be consistent to have a forcing axiom stronger than $MA(\omega_2)$, that is to generalise properness to some extent. However, the above limitations show that this generalisation cannot be straightforward. Let us review some of the known literature.

The above mentioned paper [93] by Rosłanowski and Shelah and some others in that series of articles, do obtain forcing axioms of the ilk of PFA, although these papers concentrated more on preservation by iteration theorems rather than the axiom. At the time the two were considered to go hand-in-hand, but the new proof of PFA does not go through a preservation by theorem process. In [93], Rosłanowski and Shelah introduce the notion of being *proper over semi-diamonds* and use it to define *proper over* λ. They prove an iteration theorem for proper over a fixed semi-diamond, $(< \lambda)$-closed forcing with lubs and obtain a forcing axiom for it (see Theorem 4.1 in [93].) In a different direction, David Asperó and Miguel Angel Mota in [1] obtain a forcing axiom that they call $MA^{1.5}$.

In §7.8, we have already pointed to a difference between the tree property at ω_1 and the one at ω_2. Another main difference between forcing at ω_1 and at ω_2 is the *club guessing*, a ZFC principle that was discovered by Shelah in his development of pcf theory, see [104] (we discuss pcf theory in 17). The statement of club guessing is similar to $\diamond$, but only club sets are guessed. We give the version of club guessing that is most relevant to us here, but several other versions are to be found in [104]. The proof is surprisingly simple and quite ingenious. We use the notation S^2_0 for $S^{\aleph_2}_{\aleph_0}$.

Theorem 16.6.1 (Shelah) *Say that a sequence*

$$\langle C_\delta : \delta \in S \rangle$$

for some S stationary subset of S^2_0 and such that each C_δ is an unbounded subset of the corresponding δ with $\mathrm{otp}(C_\delta) = \omega$, is a candidate. *Then there is a candidate which* guesses clubs of ω_2, *that is such that for every club C of ω_2,*

$$\{\delta : C_\delta \subseteq C\} \text{ is stationary.}$$

Proof We shall perform a construction by induction on $\zeta < \omega_1$.

At the stage 0, let $\langle C^0_\delta : \delta \in S^2_0\rangle$ be any candidate. If it guesses clubs, we are done. Otherwise, there is a club C^0 of ω_2 such that $\{\delta : C^0_\delta \subseteq C^0\}$ is non-stationary, and therefore, there is a club D^0 of ω_2 such that

$$\delta \in D_0 \cap S^2_0 \implies C^0_\delta \setminus C^0 \neq \emptyset.$$

Let $E_0 = \lim(C^0 \cap D^0)$, so still a club of ω_2. Let $S_0 = \lim(E_0) \cap S$.

Given a club E_ζ of ω_2 and a candidate $\langle C^\zeta_\delta : \delta \in S_\zeta\rangle$ which does not guess clubs of ω_2, we replace $\langle C^\zeta_\delta : \delta \in S_\zeta\rangle$ by another candidate,

$$\langle C^{\zeta+1}_\delta : \delta \in S_{\zeta+1}\rangle$$

as follows. We first find a club $C^{\zeta+1}$ of ω_2 which exemplifies that $\langle C^\zeta_\delta : \delta \in S_\zeta\rangle$ does not guess clubs, so such that there is a club $D_{\zeta+1}$ with

$$\delta \in D_{\zeta+1} \cap S_\zeta \implies C^\zeta_\delta \setminus C^{\zeta+1} \neq \emptyset.$$

Let $E_{\zeta+1} = \lim(C\zeta + 1 \cap D_{\zeta+1})$, so still a club of ω_2. Let $S_{\zeta+1} = \lim(E_{\zeta+1}) \cap S_\zeta$, so still a stationary subset of S^2_0. For $\delta \in S_{\zeta+1}$, we let

$$C^{\zeta+1}_\delta = \{\sup(\alpha \cap E_{\zeta+1}) : \alpha \in C^\zeta_\delta \setminus \min(E_{\zeta+1})\}.$$

Notice that for $\delta \in S_{\zeta+1}$, we have that $\delta \in \lim(E_{\zeta+1})$, so $\sup(C^{\zeta+1}_\delta) = \delta$, and also note that $C^{\zeta+1}_\delta \subseteq E_{\zeta+1}$.

For $\zeta \leq \omega_1$ limit ordinal let $E_\zeta = \bigcap_{\xi<\zeta} E_\xi$, still a club of ω_2. Let $S_\zeta = S \cap \lim(E_\zeta)$.

At the end of this induction, let us take any $\delta^* \in \lim(E_{\omega_1}) \cap S_{\omega_1}$. Notice that for $\alpha > \min(E_{\omega_1})$, we have that $\langle \sup(\alpha \cap E_\zeta) : \zeta < \omega_1\rangle$ is a non-increasing sequence of ordinals, so it must be eventually constant. For each such α let $\zeta_\alpha < \omega_1$ be an ordinal after which $\langle \sup(\alpha \cap E_\zeta) : \zeta < \omega_1\rangle$ is constant. Let

$$\zeta^* = \sup_{\alpha \in C_{\delta^*} \setminus \min(E_{\omega_1})} \zeta_\alpha,$$

so $\zeta^* < \omega_1$. Then we have that $C^{\zeta+2}_{\delta^*} = C^{\zeta+1}_{\delta^*} \subseteq E_{\zeta+1}$, which is in a contradiction with the choice of $\langle C^{\zeta+2}_\delta : \delta \in S_{\zeta+2}\rangle$. This contradiction shows that the induction must have stopped at some moment, so we must have found a sequence that guesses clubs. ★16.6.1

In contrast with Theorem 16.6.1, it is consistent that there is no club guessing between ω and ω_1. In particular PFA implies that there is no such guessing. This was proved by Shelah, see [VII 3.14.A , [105]].

Theorem 16.6.2 (Shelah) *PFA implies that there is no sequence $\langle C_\delta : \delta \in S\rangle$ for some S stationary subset of the set of countable limit ordinals such that* *each C_δ is an unbounded subset of the corresponding δ with* $\text{otp}(C_\delta) = \omega$, *and such that for every club C of ω_1,*

$$\{\delta : C_\delta \subseteq C\} \textit{ is stationary.}$$

Putting the two above theorems together, we see that whatever a generalisation of PFA we may decide to adopt at ω_2, it will not be able to deal with the same kind of forcing as the one we are used to having at ω_1. There are many other known obstacles of this kind, which make lifting PFA to higher cardinals a very challenging question.

Even at ω_1 there, is still much to discover in the theory of forcing axioms. A main challenge is to develop a sound theory of forcing which preserves the small value of the continuum, hopefully with a corresponding forcing axiom consistent with CH or, on the other hand, to find a forcing axiom stronger than MA but consistent with the value of the continuum equal to $\aleph_3$. Many partial results are known, but there are notable problems that resist attempts. One of them is the *Measuring* introduced by Moore, see [29]. It is a strong negation of club guessing, expressed as in the following:

Measuring. For every sequence $\langle C_\delta : \delta \text{ limit } < \omega_1\rangle$ where each C_δ is a closed unbounded subset of δ, there is a club C of ω_1 such that

- either for some $\alpha < \delta$ we have $(C \cap \delta) \setminus \alpha \subseteq C_\delta$ or
- $(C \setminus \alpha) \cap C_\delta = \emptyset$.

Notice that in either case, C_δ does not guess $\lim(C)$.

It is not clear if measuring is consistent with CH or with $2^{\aleph_0} = \aleph_3$ since published work claiming such consistencies unfortunately seems to have a gap.

17

Singular Cardinal Hypothesis and Some PCF

17.1 Easton Forcing

Not that long after the discovery of Cohen's forcing, William Easton [28] generalised Cohen's idea of dealing with the power set of one fixed regular cardinal by being able to deal with a whole class of regular cardinals simultaneously. Namely, he proved:

Theorem 17.1.1 *Suppose that $\mathcal{R}$ is a class (possibly a proper class) of regular cardinals and that F is a class function defined on $\mathcal{R}$ and with values in the class of cardinals, such that*

- *F is non-decreasing, that is if $\kappa \leq \lambda$ are two elements of $\mathcal{R}$, then $F(\kappa) \leq F(\lambda)$,*
- *for every $\kappa \in \mathcal{R}$, we have $\mathrm{cf}(F(\kappa)) > \kappa$.*

Then, if the ground model satisfies GCH, there is a cardinal and cofinality preserving forcing notion $\mathbb{P}$ such that in the extension by a $\mathbb{P}$-generic filter, we have

$$2^\kappa = F(\kappa) \text{ for all } \kappa \in \mathcal{R}.$$

Proof (outline) Easton forcing is essentially an iteration of Cohen forcing with *Easton supports*, which we now explain. The conditions $p \in \mathbb{P}$ are class functions with $\mathrm{dom}(p) = \mathrm{dom}(F)$ and such that for all $\kappa \in \mathcal{R}$, the value of $p(\kappa)$ is a condition in $\mathrm{Add}(\kappa, 2^\kappa)$. There is an additional requirement that

$$(\forall \lambda \text{ regular }) \, |\,\mathrm{supt}(p) \cap \lambda| < \lambda.$$

The order on $p \in \mathbb{P}$ is coordinatewise, that is $p \leq q$ iff

$$(\forall \kappa \in \mathrm{dom}(p))(p(\kappa) \leq q(\kappa)).$$

In fact, what we have defined here is *the Easton product* of Cohen forcing over the cardinals in $\mathcal{R}$, since for every $p \in \mathbb{P}$ and an ordinal $\kappa \in \text{dom}(p)$, we have that $p(\kappa)$ is a ground model forcing notion. The point of Easton support is that it allows the forcing to be split into a product (or an iteration, in the case of an actual iteration) of a 'small' forcing followed by a 'closed' forcing, as we now explain.

For any λ a regular cardinal let

$$\mathbb{P}_\lambda = \{p \in \mathbb{P} : \ \text{supt}(p) \subseteq \lambda\} \text{ and } \mathbb{P}_{>\lambda} = \{p \in \mathbb{P} : \ \text{supt}(p) \cap \lambda = \emptyset\},$$

both with the order inherited from $\mathbb{P}$. Then we can make the following observation:

Claim 17.1.2 Let λ be any regular cardinal.
(1) The forcing $\mathbb{P}$ is isomorphic to $\mathbb{P}_\lambda \times \mathbb{P}_{>\lambda}$, so forcing with $\mathbb{P}$ is equivalent to first forcing with $\mathbb{P}_\lambda$ and then with $\mathbb{P}_{>\lambda}$, or (more conveniently for this proof) equivalent to first forcing with $\mathbb{P}_{>\lambda}$ and then with $\mathbb{P}_\lambda$.
(2) Under the assumption $2^{<\lambda} = \lambda$, the forcing $\mathbb{P}_\lambda$ has λ^+-cc.
(3) $\mathbb{P}_{>\lambda}$ is $(< \lambda^+)$-closed.

Proof of Claim 17.1.2 We only prove part (2), as the other parts are an exercise. For the explanation of the conclusion of (1), see [VII 1.4 [63]].

For any $p \in \mathbb{P}_\lambda$, we consider $d(p) = \bigcup_{\kappa \in \text{dom}(p)} \{\kappa\} \times \text{dom}(p(\kappa))$. As $|\text{dom}(p)| < \lambda$ and λ is regular, we have $|d(p)| < \lambda$. Given $\{p_\alpha : \alpha < \lambda^+\}$ in $\mathbb{P}_\lambda$, by the assumption $2^{<\lambda} = \lambda$, we can apply the Δ-System Lemma to pass to a set A of cardinality λ^+ such that $\{d(p_\alpha) : \alpha \in A\}$ form a Δ-System with root r^*. Then, using the fact that $|r^*| < \lambda$ and $2^{<\lambda} = \lambda$, we can assume that the restrictions $p_\alpha(\zeta) \upharpoonright r_\zeta$ are all the same, for $\alpha \in A$, where $\zeta \times r_\zeta$ is $r^* \cap \{\zeta\} \times \zeta$. Then every two conditions p_α, p_β for $\alpha, \beta \in A$ are compatible, as is exemplified by their union. $\bigstar_{17.1.2}$

Lemma 17.1.3 *Easton forcing over a model of GCH preserves cardinals and cofinalities.*

Proof Suppose otherwise, so that G is $\mathbb{P}$-generic in which some regular λ has changed its cofinality to θ. By Claim 17.1.2(3) and and Lemma 9.1.2(2), a cofinal function f from θ to λ could not be added by $\mathbb{P}_{>\lambda}$, as it would have then already been present in the ground model. By Claim 17.1.2(2) and Lemma 9.1.2(1), any function g from θ to λ in the extension by $\mathbb{P}_\lambda$ satisfies $(\forall i)\, g(i) \leq h(i)$ for some ground model function h, a contradiction as then if g is cofinal in λ, so is h. $\bigstar_{17.1.3}$

It is clear that $\mathbb{P}$ forces $2^\kappa = F(\kappa)$ for all in $\kappa \in \mathcal{R}$, which finishes the proof. $\bigstar_{17.1.1}$

At this point one may wonder if the restriction on the class $\mathcal{R}$ only containing regular cardinals, is actually necessary. The answer is quite intricate since it turns out that the powers of singular cardinals are to some extent controlled by the powers of the regulars below. We shall explain that in §17.2.

17.2 Large Cardinals and Singular Cardinals

An important and surprising remark on forcing can be made by analysing Cohen forcing in the light of an earlier result of Scott (unpublished, but see [55]):

Observation 17.2.1 (Scott) *Suppose that κ is a measurable cardinal in a model of GCH and we force with* $\mathrm{Add}(\kappa, \kappa^{++})$. *Then κ is not measurable in the generic extension.*

Proof Forcing with $\mathrm{Add}(\kappa, \kappa^{++})$ preserves cardinals and GCH below κ and yet $2^\kappa > \kappa^+$ in the extension. Scott proved that a measurable cardinal cannot be the first to fail GCH, moreover if for a measurable λ we have $2^\lambda > \lambda^+$, then there is a measure one set of $\theta < \lambda$ such that $2^\theta > \theta^+$. Therefore, in our case, κ is no longer measurable in the extension. ★17.2.1

This theorem, proven using the method of ultrapowers, inspired further developments, including a very surprising one by Silver in [109]. He used a combination of the technique of elementary embeddings in combination with forcing and his theory of lifting such embeddings through generic extensions, to prove:

Theorem 17.2.2 (Silver) *A singular cardinal of uncountable cofinality cannot be the first to fail GCH, moreover, if for such a singular cardinal λ we have $2^\lambda > \lambda^+$, then there is a stationary set of $\theta < \lambda$ such that $2^\theta > \theta^+$.*

This result means that the fact that Easton's theorem 17.1.1 deals only with regular cardinals, is not a coincidence. We cannot change the values of the powers of the singular cardinals without paying attention to the value of the power set function below. This work led to a very influential hypothesis:

Singular Cardinal Hypothesis (SCH) For a singular strong limit cardinal κ, the value of 2^κ is κ^+.

Another, strictly stronger version of SCH is

$$2^{\mathrm{cf}(\kappa)} < \kappa \implies \kappa^{\mathrm{cf}(\kappa)} = \kappa^+.$$

Although one can easily change a model where the first version of SCH fails to one where it does not but it fails the second one, the essential difficulty of producing a model in which SCH fails remains the same for the versions. Therefore we shall speak loosely just of SCH in general.

The first model to satisfy the failure of SCH was produced by Silver in 1971 (unpublished), from the assumption of the existence of a supercompact cardinal. First he produced a model where $2^\kappa > \kappa^+$ holds for a measurable cardinal κ, and then proved that forcing with Prikry forcing over that model provides one where SCH fails. He used the technique of lifting the embeddings. His method is well described in [pp. 384–385 [56]].

Magidor [70] showed that the first cardinal which fails GCH can indeed be singular, necessarily of countable cofinality, and moreover, that it can be $\aleph_\omega$, namely:

Theorem 17.2.3 (Magidor) *If κ is a supercompact cardinal, then there is a forcing extension in which κ becomes $\aleph_\omega$ and strong limit, but where $2^{\aleph_\omega} > \aleph_{\omega+1}$.*

The striking difference between Silver's theorem 17.2.2 and Magidor's theorem 17.2.3 indicates that singular cardinals of cofinality ω are different from other singular cardinals, a difference that is not completely understood even now. Magidor's theorem also raised the question of what exactly is needed as a large cardinal assumption to obtain such a result. That something was needed was already known, by a celebrated theorem of Jensen. In [22] he discovered a *covering theorem* which implies that failing SCH requires some large cardinal strength. Namely, let $0^\#$ be defined as follows:

$0^\#$ states that there is a non-trivial elementary embedding $j : \mathbf{L} \to \mathbf{L}$.

The statement $0^\#$ cannot be true in $\mathbf{L}$, because of the following theorem by Kunen from [61].

Theorem 17.2.4 (Kunen) *There is no non-trivial elementary embedding $j : \mathbf{V} \to \mathbf{V}$.*

Before stating the Covering Lemma, we use the occasion to describe the difference between *small large cardinal* assumptions and the *large* ones. The statement $0^\#$ is implied by various large cardinal axioms, including for example that of a measurable cardinal (exercise). This shows that the assumption of $0^\#$ or that of the existence of a measurable cardinal are incompatible with $\mathbf{V} = \mathbf{L}$, as opposed for example to the assumption of a weakly compact cardinal which, if consistent, is perfectly compatible with $\mathbf{V} = \mathbf{L}$. The latter kind of large cardinal assumptions are called *small large cardinals* and the former, the *large large cardinals*.

The covering theorem of Jensen is the following:

Theorem 17.2.5 (Jensen) *Suppose that* $0^\sharp$ *does not exist. Then for every uncountable set* $A \in \mathbf{V}$ *there is a set* $B \in \mathbf{L}$ *with* $A \subseteq B$ *and* $|A| = |B|$.

A corollary is:

Corollary 17.2.6 *If there is no* $0^\sharp$*, then SCH holds.*

In particular, failing SCH requires large cardinal axioms. In a seminal work by Gitik [41], he proved:

Theorem 17.2.7 *The exact consistency strength of the failure of SCH is the assumption of the existence of a measurable cardinal* κ *satisfying* $o(\kappa) = \kappa^{++}$.

The assumption $o(\kappa) = \kappa^{++}$ refers to the *Mitchell* order between ultrafilters introduced in [78], which is defined for normal ($< \kappa$)-complete ultrafilters U and U' on the same measurable cardinal κ by

$$U \lhd U' \iff U \in \mathrm{Ult}(V, U').$$

It turns out that $\lhd$ is a well founded relation, so each such ultrafilter U' has a recursively defined ordinal rank $o(U') = \sup\{o(U) + 1 : U \lhd U'\}$. The measurable cardinal κ then has a Mitchell order $o(\kappa)$ which is defined by $o(\kappa) = \sup\{o(U) + 1 : U \text{ a normal ultrafilter on } \kappa\}$.

We have already mentioned that SCH fails above a strongly compact cardinal (Solovay [113]) and is implied by PFA (Viale [123]); see the discussion in §15.5.

Finally, in what is in most respects an ultimate theorem in the context of failing GCH, we may ask if and how it can fail everywhere. Foreman and Woodin in [33] showed that modulo large cardinals, it is consistent that GCH fails at every cardinal. They used *Radin forcing*, which can be used either to change cofinalities to various values of the new cofinality (Prikry forcing necessarily gives cofinality ω) or to preserve large cardinal properties, under suitable assumptions. Cummings [13] proved that GCH can hold at all successors but fail at all limit cardinals. Other variations of this question have been studied.

A compact but illuminating introduction to this topic, including a presentation of *Magidor forcing* for changing cofinalities, is in [§36, [51]]. A thorough reference is Gitik's article [42].

17.2.1 PCF Theory

Shelah's thesis about the Singular Cardinal Hypothesis has been that even though SCH can fail, it cannot fail dramatically. Indeed in the late 1980s, see the book [104], he established a celebrated theorem to that extent.

Theorem 17.2.8 (Shelah) *Suppose that $\aleph_\omega$ is a strong limit. Then $2^{\aleph_\omega} < \aleph_{\omega_4}$.*

The bound $\aleph_{\omega_4}$ is not known to be sharp and there is much research on trying to improve it. However, the point that Shelah makes is that the bound is less important than the way that it was obtained. In particular, he argues, one should no longer be asking questions about cardinal arithmetic, but questions that come from new operations on cardinals. Indeed, the above theorem is just an instance of a wealth of combinatorial results that are obtained in [104]. They are a consequence of the introduction of two new operations on cardinal numbers, the *pcf*, the possible cofinalities and the *pp*, the pseudo-power. The point is that the values of these functions at a singular cardinal are rather absolute. In some sense, the possible cofinalities are a refinement of the study of cardinal arithmetic using the ultrapowers and ultrafilters, because rather than studying that, pcf theory studies the family of objects of the kind

$$\Pi_{i<\theta}\lambda_i/F$$

where each λ_i is a regular cardinal and $\sup_{i<\theta}\lambda_i$ is some singular cardinal κ of cofinality θ. The set F is a non-principal ultrafilter on κ. Although it is initially quite surprising to a set theorist, used to the quite arbitrary choice of the ultrafilter in the arguments with elementary embeddings, to consider that changing the base ultrafilter may change anything, the view taken by pcf theory is that looking at all possible cofinalities of such a product gives important information about κ. This type of study has appeared in model theory, including a considerable amount of Shelah's own work (see his book [102]) and therefore pcf is another occasion after his invention of proper forcing where Shelah masterly recast pure model-theoretic arguments into a new and surprising context in set theory. A more recent example which strikingly solved a long open question in set theory (showing that the two cardinals $\mathfrak{p}$ and $\mathfrak{t}$ are equal) using a similar model-theoretic insight from the study of the so called Keisler order, is the work of Maryanthe Malliaris and Shelah in [72].

Unfortunately we do not have the scope to develop pcf theory right here. Let us only mention one more wonderful consequence of this theory, which is Shelah's so called *Revised GCH* from [106]. It introduced the $\lambda^{[\kappa]}$ which measures the minimal cardinality of a subfamily $\mathcal{P}$ of $[\lambda]^\kappa$ such that each element of $[\lambda]^\kappa$ is contained in the union of $< \kappa$ elements of $\mathcal{P}$.

Theorem 17.2.9 (Shelah) *Suppose that μ is a strong limit singular and $\lambda > \mu$. Then for all but boundedly many $\kappa < \mu$ we have $\lambda^{[\kappa]} = \lambda$.*

The introduction to [106] contains much motivation and the explanation why the above statement should be called Revised GCH. Several different proofs of this theorem are available in the literature. A much simplified proof appears in a later paper by Shelah [107].

A main question in this subject, which seems to have been solved in the negative in an as of yet unverified preprint by Gitik, is the following:

Question 17.2.10 Suppose that $\mathfrak{a}$ is a set of regular cardinals with $\min(\mathfrak{a}) > |\mathfrak{a}|$. Is then $|\mathrm{pcf}(\mathfrak{a})| = |\mathfrak{a}|$?

18
Forcing at Singular Cardinals and Their Successors

The last sections of Chapter 17 show why the forcing techniques that we have at $\aleph_1$ cannot possibly work at singular cardinals. Similar concerns apply to the successors of singular cardinals. For example, if we wish to study values of various cardinal invariants at κ, then we had better make sure that they are not trivially equal to κ^+. Therefore we wish to work in the context of $2^\kappa > \kappa^+$. If, in addition, we have that κ is a strong limit cardinal, then we are automatically dealing with the failure of SCH and so with large cardinals. In this situation, the cardinal invariants at κ^+ are also affected. This situation presents many challenges and at this moment there is no unique technique or an axiom that deals with it. However, some techniques have emerged in recent years, two of which will be described below.

18.1 Unions of Ultrafilters

Here is a typical difficulty in the context of forcing at singular cardinals. In order to fail SCH at κ, we typically start with κ being some large cardinal. To make matters simple, let us assume that we have a supercompact cardinal and that we have prepared it using the Laver preparation. This gives us a chance to do an iteration $\mathbb{P}$ of a $(< \kappa)$-directed closed forcing that will let us take care of some combinatorial fact that we wish to deal with at the end and keep the supercompactness in reserve, so that at the very end we shall change the cofinality by using some Prikry-like forcing $\mathbb{Q}$. The point though is that we have to stop after changing the cofinality, since then we have a singular cardinal and we cannot use the usual techniques of forcing to deal with it. Let us say that the final universe is $V[G][H]$ where G is generic for the iteration $\mathbb{P}$ and H for the Prikry-like forcing $\mathbb{Q}$. If we wish to predict the behaviour of $V[G][H]$ during the iteration that produces G, we need to predict the behaviour not just of

$\mathbb{P}$-names as in the usual bookkeeping, but of $\mathbb{P}$-names for $\mathbb{Q}$-names! Furthermore, if we wish to end up in the context where we fail SCH, we need to be adding subsets of κ during the iteration $\mathbb{P}$. But then, if $\mathbb{Q}$ is for example, Prikry forcing, then we need to have the access to the ultrafilter that will be used to do this forcing. Since we are adding subsets to κ as we go in $\mathbb{P}$, no ultrafilter on κ that we shall see on the way will survive as an ultrafilter by the time we get to $\mathbb{Q}$.

A technique that has proven interesting in this context is to force 'two steps' in advance and to build the ultrafilter as we go. It was discovered by Gitik and Shelah in [43], where they studied densities of box products. The ultrafilter $\mathcal{D}$ is built as the union of filters $\mathcal{F}$ which were ultrafilters at some stage of the iteration. The point is that in this way the $\text{Pr}(\mathcal{F})$-names act as a *prediction* for the $\text{Pr}(\mathcal{D})$-names at the end of the iteration, so we can control them along the iteration itself. A similar technique adapted to other kinds of situations, such as Radin forcing or Prikry forcing with interleaved collapses was used by Džamonja and Shelah in [25] who dealt with small universal families of graphs at $\kappa+$ where κ is a singular cardinal of countable cofinality, in Cummings, Džamonja, Magidor, Morgan and Shelah in [17] for the same problem where κ is a singular cardinal of arbitrary cofinality, by Cummings, Džamonja and Morgan in [16] for $\kappa = \aleph_\omega$ and by Jacob Davis [20] for $\aleph_{\omega_1+1}$. A typical result of these papers resembles the main theorem of [17], which is given below.

Theorem 18.1.1 (Cummings, Džamonja, Magidor, Morgan and Shelah) *Suppose κ is a supercompact cardinal, $\lambda < \kappa$ is a regular cardinal and θ is a cardinal with $\text{cf}(\theta) \geq \kappa^{++}$ and $\kappa^{+3} \leq \theta$. Then there is a forcing extension in which cofinally many cardinals below κ, κ itself and all cardinals greater than κ are preserved, $\text{cf}(\kappa) = \lambda$, $2^\kappa = 2^{\kappa^+} = \theta$ and there is a universal family of graphs on κ^+ of size κ^{++}.*

The idea of building a good ultrafilter as a union of filters originated in earlier work of Kunen in [62], who built what is literally called a *good* ultrafilter over a given cardinal λ just in ZFC using that technique, while such ultrafilters were known to exist only under the assumption $2^\lambda = \lambda^+$. What is new in the above described is the prediction of names in the Prikry-like forcings.

The technique has also been used for some problems regarding the chromatic number of graphs, by Džamonja, Péter Komjáth and Morgan in [27] and to discuss the cardinal invariants of singular cardinals, as we now explain. Clearly, an interesting invariant to consider in this context is the number $\mathfrak{u}(\kappa)$, which is the minimal size of a base of a uniform ultrafilter on κ. Here, a *base* of an ultrafilter $\mathcal{U}$ on κ is a family $\mathcal{B}$ of sets in the ultrafilter such that for every $A \in \mathcal{U}$ there is $B \in \mathcal{B}$ with $|A \setminus B| < \kappa$. An ultrafilter on κ is *uniform* if all its elements have size κ. The process of building ultrafilters as unions of filters gives a handle on calculating $\mathfrak{u}$. This investigation was instigated by Shimon Garti and Shelah in [37], who proved that from a supercompact cardinal it follows that it is consistent to have a strong limit singular cardinal κ of countable cofinality such that $\mathfrak{u}(\kappa) < 2^\kappa$. A similar theorem but for $\kappa = \aleph_\omega$ is proved in a joint work between Garti, Gitik and Shelah [38], where the optimal consistency strength of $o(\kappa) = \kappa^{++}$ is obtained. Andrew Brooke-Taylor, Vera Fischer, (Sy) Friedman and Diana Montoya [9] proved that in the type of models as in [25], many cardinal invariants at κ have the value κ^+, including the $\mathfrak{u}(\kappa)$ number. Dilip Raghavan and Shelah in [90] consider $\mathfrak{u}(\kappa)$ for various values of κ, to obtain that it is consistent, again from the assumption of the existence of a supercompact cardinal, that $\mathfrak{u}(\aleph_{\omega+1}) < 2^{\aleph_{\omega+1}}$.

18.2 Tree Property at $\aleph_{\omega+1}$

In §7.8 we have discussed the existence of λ-Aronszajn trees for various λ. Such a tree is an instance of an incompactness phenomenon, as is the principle $\Box_\lambda$, since the former does not have an unbounded branch and the latter does not have a club which threads all the elements of the $\Box$-sequence. Since the result of Solovay in [113] discussed in [62] we know that above a large enough cardinal there are compactness phenomena, notably $\Box_\lambda$ fails above a strongly compact cardinal. For the κ^+-Aronszajn tree even less is needed: by Specker's result ([115], see §7.8), we know that if $\kappa^{<\kappa} = \kappa$, there is a κ^+-Aronszajn tree which is a union of κ antichains, so certainly there is such κ^+-Aronszajn tree for any strongly inaccessible κ. Knowing that classical models of the failure of SCH are constructed by changing the cofinality of a large cardinal to become singular, it is interesting to ask if the compactness phenomena carry down from the large cardinal to the singular cardinal. For instances of this question regarding squares one can see the excellent paper [15] by Cummings, Foreman and Magidor, giving further references to other parts of the body of work they devoted to the connections between squares, reflection and pcf theory.

Specifically for trees, the topic of this section, in the 1980s Woodin and others asked if the failure of SCH at a singular κ of countable cofinality implies the existence of a κ^+-Aronszajn tree, see [§2 of [32]]. It is easy to see that no classical cofinality changing extension using Prikry-like forcing will destroy the κ^+-Aronszajn tree that exists while κ is a large cardinal, since a special κ^+-Aronszajn tree will remain so in any extension that preserves cardinals. There have been other more sophisticated ways to obtain the failure of SCH but for one reason or another, they do not destroy the κ^+-Aronszajn tree in the extension. A review is in the introduction to Neeman's paper [84]. To summarise, after many tries and connections that were made with pcf theory, it was fully expected that in fact the failure of SCH at κ would imply the existence of a κ^+-Aronszajn tree. Magidor and Shelah in [71] proved the impressive ZFC result that if λ is a singular limit of strongly compact cardinals, then λ^+ has the tree property. Furthermore they, using large cardinal properties close to a huge cardinal, were able to obtain the tree property at $\aleph_{\omega+1}$. Nobody knew how to bring that down to the successor of a singular cardinal failing SCH. Neeman's main result in [84] was to do exactly that, using a different method of proof. This showed that the prediction that the failure of SCH at κ implies the existence of a κ^+-Aronszajn tree was wrong.

Theorem 18.2.1 (Neeman) *Suppose that there is a sequence of ω supercompact cardinals. Then there is a forcing extension in which there is a cardinal κ satisfying:*

- *κ is a strong limit and* $\mathrm{cf}(\kappa) = \omega$,
- *SCH fails at κ, specifically $2^\kappa = \kappa^{++}$ and*
- *there are no κ^+-Aronszajn trees.*

Neeman's method is out of the scope of this book. It has been continued and improved by Neeman and others, especially Sinapova who has contributed ideas to it from the very beginning and who has a rich research programme on the subject. See, for example, her paper [111] where starting from a sequence of ω supercompact cardinals and a weakly compact above them, she produces a generic extension in which the tree property holds at the first and second successor of a strong limit singular cardinal. Note that Neeman's result does not give $\kappa = \aleph_\omega$ and it is an open problem to obtain the statement of Theorem 18.2.1 with $\kappa = \aleph_\omega$. Sinapova [110] did obtain a model with the failure of SCH at $\aleph_{\omega^2}$ and the tree property at $\aleph_{\omega^2+1}$, from the same large cardinal assumptions as in Theorem 18.2.1.

References

[1] Asperó, David, and Mota, Miguel Angel. 2015. A generalization of Martin's axiom. *Israel J. Math*, **210**, 193–231.

[2] Avraham, Uri, and Shelah, Saharon. 1981. Martin's Axiom does not imply that every two $\aleph_1$-dense sets of reals are isomorphic. *Israel J. Math*, **38**(1-2), 161–176.

[3] Bartoszyński, Tomek, and Judah, Haim. 1995. *Set Theory: On the Structure of the Real Line*. A.K. Peters, Ltd.

[4] Baumgartner, James E. 1973. All $\aleph_1$-dense sets of reals can be isomorphic. *Fund. Math*, **79**(2), 101–106.

[5] Baumgartner, James E. 1983. Iterated forcing. Pages 1–59 of: Mathias, Adrian R.D. (ed), *Surveys in Set Theory*. London Math. Soc. Lecture Note Ser., vol. 87. Cambridge University Press.

[6] Baumgartner, James E. 1984. Applications of the proper forcing axiom. Pages 913–959 of: Kunen, Kenneth, and Vaughan, Jerry E. (eds), *Handbook of Set-Theoretic Topology*. Amsterdam: North-Holland.

[7] Baumgartner, James E., Malitz, Jerome, and Reinhardt, William. 1970. Embedding Trees in the Rationals. *Proceedings of the National Academy of Sciences United States of America*, **67**(4), 1748–1753.

[8] Borel, Émile. 1919. Sur la classification des ensembles de mesure nulle. *Bull. Soc. Math. France*, **47**, 97–125.

[9] Brooke-Taylor, Andrew D., Fischer, Vera, Friedman, Sy D., and Montoya, Diana C. 2017. Cardinal characteristics at κ in a small $\mathfrak{u}(\kappa)$ model. *Ann. Pure Appl. Logic*, **168**(1), 37–49.

[10] Burali-Forti, Cesare. 1897. Una questione sui numeri transfiniti, sulle classi ben ordinate. *Rend. Circ. Mat. Palermo*, **11**(154-164; 260).

[11] Cantor, Georg. 1874. Ueber eine Eigenschaft des Inbegriffes aller reellen algebraischen Zahlen. *J. Reine Angew. Math.*, **77**, 258–262.

[12] Cohen, Paul. 1963. The independence of the Continuum Hypothesis. *Proc. Natl. Acad. Sci. USA*, **50**(6), 1143–1148.

[13] Cummings, James. 1992. A model in which GCH holds at successors but fails at limits. *Trans. Amer. Math. Soc.*, **329**(1), 1–39.

[14] Cummings, James. 2010. Iterated forcing and elementary embeddings. Pages 775–883 of: *Handbook of Set Theory. Vols. 1, 2, 3*. Springer, Dordrecht.

[15] Cummings, James, Foreman, Matthew, and Magidor, Menachem. 2001. Scales, squares and reflection. *J. Math. Log.*, **1**(1), 35–98.

[16] Cummings, James, Džamonja, Mirna, and Morgan, Charles. 2016. Small universal families of graphs on $\aleph_{\omega+1}$. *J. Symb. Log.*, **81**(2), 541–569.

[17] Cummings, James, Džamonja, Mirna, Magidor, Menachem, Morgan, Charles, and Shelah, Saharon. 2017. A framework for forcing constructions at successors of singular cardinals. *Trans. Amer. Math. Soc.*, **369**(10), 7405–7441.

[18] Cummings, James, Džamonja, Mirna, and Neeman, Itay. 2018 (November). *Strongly κ^+-cc forcing*. arXiv:1811.05426.

[19] Cummings, James, Hayut, Yair, Magidor, Menachem, Neeman, Itay, Sinapova, Dima, and Unger, Spencer. 2020. The ineffable tree property and failure of the singular cardinal hypothesis. *Trans. Amer. Math. Soc.*, **373**, 5937–5955.

[20] Davis, Jacob. 2017. Universal graphs at $\aleph_{\omega+1}$. *Ann. Pure Appl. Logic*, **168**(10), 1878–1901.

[21] Devlin, Keith J. 1983. The Yorkshireman's guide to proper forcing. Pages 60–115 of: Mathias, A. R. D. (ed), *Surveys in Set Theory*. London Mathematical Society Lecture Note Series. Cambridge University Press.

[22] Devlin, Keith J., and Jensen, R. B. 1975. Marginalia to a theorem of Silver. Pages 115–142. Lecture Notes in Math., Vol. 499 of: ⊨*ISILC Logic Conference (Proc. Internat. Summer Inst. and Logic Colloq., Kiel, 1974)*. Springer, Berlin.

[23] Dolinar, Gregor, and Džamonja, Mirna. 2013. Forcing Square$_{\omega_1}$ with finite conditions. *Ann. Pure Appl. Logic*, **164**, 49–64.

[24] Džamonja, Mirna. 2017. *Théorie des ensembles pour les philosophes*. Sarrebruck: Éd. Univér. Européenne.

[25] Džamonja, Mirna, and Shelah, Saharon. 2003. Universal graphs at the successor of a singular cardinal. *J. Symb. Log.*, **68**, 366–387.

[26] Džamonja, Mirna, and Shelah, Saharon. 2020. On wide Aronszajn trees in the presence of MA. *J. Symb. Log.* to appear.

[27] Džamonja, Mirna, Komjáth, Péter, and Morgan, Charles. 2004. Wild edge colourings of graphs. *J. Symb. Log.*, **69**(1), 255–264.

[28] Easton, William B. 1970. Powers of regular cardinals. *Ann. Math. Logic*, **1**, 139–178.

[29] Eisworth, Todd, Milovitch, David, and Moore, Justin Tatch. 2013. Iterated forcing and the Continuum Hypothesis. Pages 207–244 of: Cummings, James, and Schimmerling, Ernst (eds), *Appalachian Set Theory 2006–2012*. London Mathematical Society Lecture Note Series. New York: Cambridge University Press.

[30] Erdös, Paul, Hajnal, András, Mate, Atilla, and Rado, Richard. 1984. *Combinatorial Set Theory: Partition Relations for Cardinals*. Studies in Logic and the Foundations of Mathematics 106. Elsevier Science Ltd.

[31] Fodor, Géza. 1956. Eine Bemerkung zur Theorie der regressiven Funktionen. *Acta Sci. Math. (Szeged)*, **17**, 139–142.

[32] Foreman, Matthew. 2005. Some problems in singular cardinals combinatorics. *Notre Dame J. Form. Log.*, **46**, 309–322.

[33] Foreman, Matthew, and Woodin, W. Hugh. 1991. The Generalized Continuum Hypothesis can fail everywhere. *Ann. of Math. (2)*, **133**(1), 1–35.

[34] Foreman, Matthew, Magidor, Menachem, and Shelah, Saharon. 1988. Martin's Maximum, saturated ideals, and non-regular ultrafilters. I. *Ann. of Math. (2)*, **127**(1), 1–47.

[35] Fremlin, David H. 1984. *Consequences of Martin's Axiom*. Cambridge Tracts in Mathematics. Cambridge University Press.

[36] Friedman, Sy David. 2006. Forcing with finite conditions. Pages 285–296 of: Bagaria, Joan, and Todorčević, Stevo (eds), *Set Theory: Centre de Recerca Matemàtica, Barcelona 2003–04*. Trends in Mathematics. Basel: Birkhäuser Verlag.

[37] Garti, Shimon, and Shelah, Saharon. 2012. The ultrafilter number for singular cardinals. *Acta Math. Hungar.*, **137**(4), 296–301.

[38] Garti, Shimon, Gitik, Moti, and Shelah, Saharon. 2020. Cardinal charatecristics at $\aleph_\omega$. *Acta Math. Hungar.*, **160**, 320–336.

[39] Geschke, Stefan, and Quickert, Sandra. 2000. On Sacks forcing and the Sacks property. Pages 1–49 of: Löwe, Benedikt, Malzkorn, Wolfgang, and Räsch, Thoralf (eds), *Foundations of the Formal Science*. Applications of Mathematical Logic in Philosophy and Linguistics, vol. II. Bonn: Springer.

[40] Gilton, Thomas, and Neeman, Itay. 2017. Side conditions and iteration theorems. In: *Proceeeding of the 9th Appalachian Set Theory Workshop*.

[41] Gitik, Moti. 1991. The strength of the failure of the singular cardinal hypothesis. *Ann. Pure Appl. Logic*, **51**(3), 215–240.

[42] Gitik, Moti. 2010. Prikry-type forcings. Pages 1351–1447 of: Foreman, Matthew, and Kanamori, Akihiro (eds), *Handbook of Set Theory*, vol. 2. Dordrecht , Heidelberg, London, New York: Springer.

[43] Gitik, Moti, and Shelah, Saharon. 1998. On densities of box products. *Topology Appl.*, **88**(3), 219–237.

[44] Gödel, Kurt. 1930. Die Vollständigkeit der Axiome des logischen Funktionenkalküls. *Monatsh. Math. Phys.*, **37**(1), 349–360.

[45] Gödel, Kurt. 1931. Über formal unentscheidbare Sätze der Principia Mathematica und verwandter Systeme, I. *Monatsh. Math. Phys.*, **38**, 173–198.

[46] Gödel, Kurt. 1940. *The Consistency of the Continuum-Hypothesis*. Princeton University Press.

[47] Goldstern, Martin. 1993. Tools for your forcing construction. Pages 305–360 of: Judah, Haim (ed), *Set Theory of the Reals*. Israel Mathematical Conference Proceedings, vol. 6.

[48] Halbeisen, Lorenz J. 2012. *Combinatorial Set Theory, with a Gentle Introduction to Forcing*. Springer Monographs in Mathematics. London: Springer.

[49] Hausdorff, Felix. 1908. Grundzüge einer Theorie der geordneten Mengen. *Math. Ann.*, **65**, 435–505.

[50] Jech, Thomas. 1986. *Multiple Forcing*. Cambridge University Press.

[51] Jech, Thomas. 2003. *Set Theory*. 3rd millennium edn. Berlin Heidelberg: Springer-Verlag.

[52] Jech, Thomas. 2010. Stationary sets. Pages 93–128 of: *Handbook of Set Theory. Vols. 1, 2, 3*. Dordrecht, Heidelberg, London, New York: Springer.

[53] Jech, Thomas J. 1971. The closed unbounded filter over $\mathcal{P}_\kappa(\lambda)$. *Notices Amer. Math. Soc.*, **18**, 663.

[54] Jensen, Ronald Björn. 1972. The fine structure of the constructible hierarchy. *Ann. Math. Logic*, **4**, 229–308; erratum, ibid. 4 (1972), 443. With a section by Jack Silver.

[55] Kanamori, Akihiro. 2003. *The Higher Infinite*. Second edn. Springer Monographs in Mathematics. Berlin: Springer-Verlag.

[56] Kanamori, Akihiro. 2012. Large Cardinals with Forcing. Pages 359–414 of: Gabbay, Dov M., Kanamori, Akihiro, and Woods, John (eds), *Sets and Extensions in the Twentieth Century*. Handbook of the History of Logic, vol. 6. North Holland.

[57] Keisler, H. J., and Tarski, A. 1964. From accessible to inaccessible cardinals. Results holding for all accessible cardinal numbers and the problem of their extension to inaccessible ones. *Fund. Math.*, **53**, 225–308.

[58] König, Jules. 1906. Sur la théorie des ensembles. *Compt. rend. hebdo. Acad. France*, **143**, 110–112.

[59] Koszmider, Piotr. 2000. On strong chains of uncountable functions. *Israel J. Math.*, **118**, 289–315.

[60] Krueger, John. 2014. Coherent adequate sets and forcing square. *Fund. Math.*, **224**(3), 279–300.

[61] Kunen, Kenneth. 1971. Elementary embeddings and infinitary combinatorics. *J. Symb. Log.*, **36**(3), 407–413.

[62] Kunen, Kenneth. 1972. Ultrafilters and independent sets. *Trans. Amer. Math. Soc.*, **172**, 299–306.

[63] Kunen, Kenneth. 1980. *Set Theory*. Studies in Logic and the Foundations of Mathematics, vol. 102. North-Holland.

[64] Kurepa, Georges (Giuro). 1935. *Ensembles ordonnés et ramifiés*. Ph.D. thesis, Université de Paris, France.

[65] Laver, Richard. 1976. On the consistency of Borel's conjecture. *Acta Math.*, **137**, 151–169.

[66] Laver, Richard. 1978. Making the supercompactness of κ indestructible under κ-directed closed forcing. *Israel J. Math*, **29**(4), 385–388.

[67] Laver, Richard, and Shelah, Saharon. 1981. The $\aleph_2$-Souslin Hypothesis. *Trans. Amer. Math. Soc.*, **264**(2), 411–417.

[68] Lebesgue, Henri. 1902. Intégrale, longeur, aire. *Ann. Mat. Pura Appl.*, **7**(3), 231–359.

[69] Lévy, Azriel. 1963. Independence results in set theory by Cohen's method. IV. *Notices Amer. Math. Soc.*, **10**, 593.

[70] Magidor, Menachem. 1977. On the singular cardinals problem. I. *Israel J. Math.*, **28**(1-2), 1–31.

[71] Magidor, Menachem, and Shelah, Saharon. 1996. The tree property at successors of singular cardinals. *Arch. Math. Logic*, **35**(5-6), 385–404.

[72] Malliaris, Marianthe, and Shelah, Saharon. 2016. Cofinality spectrum theorems in model theory, set theory, and general topology. *J. Amer. Math. Soc.*, **29**, 237–297.

[73] Mathias, Adrian R. D. 1977. Happy families. *Ann. Math. Logic*, **12**(1), 59–111.

[74] Mekler, A. 1990. Universal structures in power $\aleph_1$. *J. Symb. Log.*, **55**(2), 466–477.

[75] Mekler, Alan, and Väänänen, Jouko. 1993. Trees and Π^1_1-subsets of ${}^{\omega_1}\omega_1$. *J. Symb. Log.*, **58(3)**, 1052–1070.

[76] Mekler, Alan H. 1984. c.c.c. forcing without combinatorics. *J. Symb. Log.*, **49**(3), 830–832.

[77] Mitchell, William. 1972. Aronszajn trees and the independence of the transfer property. *Ann. Math. Logic*, **5**, 21–46.

[78] Mitchell, William J. 1974. Sets constructible from sequences of ultrafilters. *J. Symb. Log.*, **39**(57-66).

[79] Mitchell, William J. 2005. Adding closed unbounded subsets of ω_2 with finite forcing. *Notre Dame J. Formal Log.*, **46**(3), 357–371.

[80] Mitchell, William J. 2009. $I[\omega_2]$ can be the nonstationary ideal on $\mathrm{Cof}(\omega_1)$. *Trans. Amer. Math. Soc.*, **361**(2), 561–601.

[81] Montague, Richard. 1957. *Contributions to the axiomatic foundations of set theory*. Ph.D. thesis, Berkeley University, USA.

[82] Moore, Justin Tatch. 2005. Set mapping reflection. *J. Math. Log.*, **5**, 87–97.

[83] Mostowski, Andrzej. 1949. An undecidable arithmetical statement. *Fund. Math.*, **36**, 143–164.

[84] Neeman, Itay. 2009. Aronszajn trees and failure of the singular cardinal hypothesis. *J. Math. Log.*, **9**, 139–157.

[85] Neeman, Itay. 2014. Forcing with sequences of models of two types. *Notre Dame J. Form. Log.*, **55**, 265–298.

[86] Ostaszewski, Adam. 1976. On countably compact, perfectly normal spaces. *J. Lond. Math. Soc.*, **14**, 505–516.

[87] Oxtoby, John C. 1980. *Measure and Category*. 2nd edn. New York: Springer-Verlag.

[88] Prikry, Karel L. 1970. Changing measurable into accessible cardinals. *Dissertationes Math. (Rozprawy Mat.)*, **68**, 5–52.

[89] Primavesi, Alexander. 2011. *Guessing axioms, invariance and Suslin trees*. Ph.D. thesis, University of East Anglia, Norwich, UK.

[90] Raghavan, Dilip, and Shelah, Saharon. 2020. A small ultrafilter number at smaller cardinals. *Arch. Math. Logic*, **53**, 325–334.

[91] Rinot, Assaf. 2011. Jensen's diamond principle and its relatives. Pages 125–156 of: *Set Theory and its Applications*. Contemp. Math., vol. 533. Providence, Rhode Island: Amer. Math. Soc.

[92] Rosenstein, Joseph R. 1982. *Linear Orderings*. Academic Press.

[93] Rosłanowski, Andrzej, and Shelah, Saharon. 2001. Iteration of λ-complete forcing notions not collapsing λ^+. *Int. J. Math. Math. Sci.*, **28**, 63–82.

[94] Rowbottom, Frederick. 1971. Some strong axioms of infinity incompatible with the axiom of constructibility. *Ann. Math. Logic*, **3**, 1–44.

[95] Russell, Bertrand. 1902 (June). *A letter to Frege*.

[96] Sacks, Gerald E. 1971. Forcing with perfect closed sets. Pages 331–355 of: Scott, Dana (ed), *Axiomatic Set Theory*. Symposia in Pure Mathematics, vol. 13. Providence, Rhode Island: Amer. Math. Soc.

[97] Scott, Dana. 1955. Definition by abstraction in axiomatic set theory. *Bull. Amer. Math. Soc.*, **61**(5), 442.

[98] Shanin, Nikolai A. 1946. A theorem for the general theory of sets. *C. R. Acad. Sci. URSS (N.S) (Doklady)*, **53**, 399–400.

[99] Shelah, Saharon. 1978. A weak generalization of MA to higher cardinals. *Israel J. Math.*, **30**(4), 297–306.

[100] Shelah, Saharon. 1980. Independence results. *J. Symb. Log.*, **45**(3), 563–573.

[101] Shelah, Saharon. 1984. Can you take Solovay's inaccessible away? *Israel J. Math.*, **48**(1), 1–47.

[102] Shelah, Saharon. 1990a. *Classification Theory and the Number of Nonisomorphic Models*. 2nd edn. Studies in Logic and the Foundations of Mathematics, vol. 92. Amsterdam: North-Holland.

[103] Shelah, Saharon. 1990b. Universal graphs without instances of CH: revisited. *Israel J. Math.*, **70**(1), 69 – 81.

[104] Shelah, Saharon. 1994. *Cardinal Arithmetic*. Oxford Logic Guides, vol. 29. New York: The Clarendon Press Oxford University Press. Oxford Science Publications.

[105] Shelah, Saharon. 1998. *Proper and Improper Forcing*. 2nd edn. Perspectives in Mathematical Logic. Berlin: Springer-Verlag.

[106] Shelah, Saharon. 2000. The Generalised Continuum Hypothesis revisited. *Israel J. Math.*, **116**, 285–321.

[107] Shelah, Saharon. 2006. More on the revised GCH and the black box. *Ann. Pure Appl. Logic*, **140**, 133–160.

[108] Silver, Jack. 1971. *Notes on reverse Easton forcing*. unpublished manuscript.

[109] Silver, Jack. 1975. On the singular cardinals problem. Pages 265–268 of: *Proceedings of the International Congress of Mathematicians (Vancouver, B. C., 1974), Vol. 1.* Canad. Math. Congress, Montreal, Que.

[110] Sinapova, Dima. 2012. The tree property and the failure of the singular cardinal hypothesis at $\aleph_{\omega^2}$. *J. Symb. Log.*, **77**(3), 934–946.

[111] Sinapova, Dima. 2016. The tree property at the first and double successors of a singular. *Israel J. Math.*, **216**(2), 799–810.

[112] Solovay, Robert M. 1970. A model of set-theory in which every set of reals is Lebesgue measurable. *Ann. of Math. (2)*, **92**, 1–56.

[113] Solovay, Robert M. 1974. Strongly compact cardinals and the GCH. Pages 365–372 of: Henkin, Leon, Addison, John, Craig, William, Scott, Dana, and Vaught, Robert (eds), *Proceedings of the Tarski Symposium, Proceedings of Symposia in Pure Mathematics, Vol. XXV*. Providence, Rhode Island: Published for the Ass. Symb. Log. by the Amer. Math. Soc.

[114] Solovay, Robert M., and Tennenbaum, Stanley. 1971. Iterated Cohen extensions and Souslin's problem. *Ann. of Math.*, **94**(2), 201–245.

[115] Specker, Ernst. 1951. Sur un problème de Sikorski. *Colloq. Math.*, **2**, 9–12.

[116] Suslin, Mikhail. 1920. Problème 3. *Fund. Math.*, **1**, 223.

[117] Todorčević, Stevo. 1984. A note on the Proper Forcing Axiom. Pages 209–218 of: *Axiomatic Set Theory (Boulder, Colorado 1983)*. Contemporary Mathematics, vol. 31. Providence, Rhode Island: Amer. Math. Soc.

[118] Todorčević, Stevo. 1989. *Partition Problems in Topology*. Providence, Rhode Island: Amer. Math. Soc.

[119] Todorčević, Stevo. 2010. Handbook of Set Theory. vol. 1. Dordrecht, Heidelberg, London, New York: Springer.

[120] Ulam, Stanisław M. 1930. Zur Masstheorie in der allgemeinen Mengenlehre. *Fund. Math.*, **16**, 140–150.

[121] Veličković, Boban. 1992. Forcing axioms and stationary sets. *Adv. Math.*, **94**(2), 256–284.

[122] Veličković, Boban, and Venturi, Giorgio. 2012. Proper forcing remastered. In: Cummings, James, and Schimmerling, Ernst (eds), *Appalachian Set Theory*. LMS Lecture Notes Series. Cambridge University Press.

[123] Viale, Matteo. 2006. The proper forcing axiom and the singular cardinal hypothesis. *J. Symb. Log.*, **71**(2), 473–479.

[124] Vitali, Giuseppe. 1905. *Sul problema della misura dei gruppi di punti di una retta*. Bologna,. Tip. Gamberini e Parmeggiani.

[125] Wagon, Stanley. 1993. *The Banach-Tarski Paradox*. Cambridge University Press.

[126] Weiß, Christoph, and Viale, Matteo. 2011. On the consistency strength of the proper forcing axiom. *Adv. Math.*, **228**(5), 2672–2687.

[127] Zermelo, Ernst. 1908. Untersuchungen über die Grundlagen der Mengenlehre. I. *Math. Ann.*, **65**(2), 261–281.

[128] Zermelo, Ernst, and Fraenkel, Adolf (Abraham). 1932. *George Cantor Gesamelte Abhandlungen*. Berlin: Julius Springer.

Index

Printed in the United States
By Bookmasters